COMPACT *Research*

World

Energy

Crisis

Current Issues

Other books in the Compact Research series include:

Drugs
> Heroin
> Marijuana
> Methamphetamine
> Nicotine and Tobacco
> Performance-Enhancing Drugs

Current Issues
> Biomedical Ethics
> The Death Penalty
> Gun Control
> Illegal Immigration

COMPACT *Research*

World

Energy

Crisis

by Stuart A. Kallen

Current Issues

ReferencePoint
Press™

San Diego, CA

© 2007 ReferencePoint Press, Inc.

For more information, contact
ReferencePoint Press, Inc.
PO Box 27779
San Diego, CA 92198
www. ReferencePointPress.com

Picture Credits:
AP/Wide World Photos, 11
Steve Zmina, 31–34, 47–50, 63–65, 78–81

Series design:
Tamia Dowlatabadi

LIBRARY OF CONGRESS CATALOGING-IN-PUBLICATION DATA

Kallen, Stuart A.
  World energy crisis / by Stuart A. Kallen.
    p. cm. — (Compact research series)
  Includes bibliographical references and index.
  ISBN-13: 978-1-60152-011-1 (hardback)
  ISBN-10: 1-60152-011-5 (hardback)
  1. Power resources. 2. Energy policy. 3. Sustainable development. 4. Renewable natural resources. I. Title.
  HD9502.A2K343 2006
  333.79'11—dc22

                                                                    2006032875

Contents

Foreword

66 Where is the knowledge we have lost in information? 99

—"The Rock," T.S. Eliot.

As modern civilization continues to evolve, its ability to create, store, distribute, and access information expands exponentially. The explosion of information from all media continues to increase at a phenomenal rate. By 2020 some experts predict the worldwide information base will double every 73 days. While access to diverse sources of information and perspectives is paramount to any democratic society, information alone cannot help people gain knowledge and understanding. Information must be organized and presented clearly and succinctly in order to be understood. The challenge in the digital age becomes not the creation of information, but how best to sort, organize, enhance, and present information.

ReferencePoint Press developed the Compact Research series with this challenge of the information age in mind. More than any other subject area today, researching current events can yield vast, diverse, and unqualified information that can intimidate and overwhelm even the most advanced and motivated researcher. The Compact Research series offers a compact, relevant, intelligent, and conveniently organized collection of information covering a variety of current and controversial topics ranging from illegal immigration to marijuana.

The series focuses on three types of information: objective single-author narratives, opinion-based primary source quotations, and facts and statistics. The clearly written objective narratives provide context and reliable background information. Primary source quotes are carefully selected and cited, exposing the reader to differing points of view. And facts and statistics sections aid the reader in evaluating perspectives. Presenting these key types of information creates a richer, more balanced learning experience.

For better understanding and convenience, the series enhances information by organizing it into narrower topics and adding design features that make it easy for a reader to identify desired content. For example, in *Compact Research: Illegal Immigration*, a chapter covering the economic impact of illegal immigration has an objective narrative explaining the various ways the economy is impacted, a balanced section of numerous primary source quotes on the topic, followed by facts and full-color illustrations to encourage evaluation of contrasting perspectives.

The ancient Roman philosopher Lucius Annaeus Seneca wrote, "It is quality rather than quantity that matters." More than just a collection of content, the Compact Research series is simply committed to creating, finding, organizing, and presenting the most relevant and appropriate amount of information on a current topic in a user-friendly style that invites, intrigues, and fosters understanding.

World Energy Crisis at a Glance

Energy Crisis

Some experts say a crisis awaits because the world's oil will be depleted by 2035 while others believe new technology and alternative energy sources will avert a crisis.

Worldwide oil consumption was about 84 million barrels a day in 2005, and demand is expected to increase by 2 million barrels a day in the coming years.

Natural Gas Consumption

Natural gas is used to produce about 40 percent of all electricity worldwide, and demand has been growing at 2 percent annually for the last 20 years.

Natural Gas Shortages

Many natural gas fields are in decline, including those in Great Britain and the United States.

Coal for Fuel

Coal is a major energy source, but converting it to gas or liquid fuel is expensive, and using it as a replacement for gasoline would require mining on a scale never yet attempted.

Nuclear Power

Nuclear power does not contribute to global warming or produce air pollution, but there is only about a 50-year supply of uranium left at current usage.

Alternative Energy

Solar cells, wind turbines, and hybrid cars are proven alternative energy technologies already in use, but no renewable energy source known today can meet the massive power demands of modern society.

Overview

66 It took us 125 years to use the first trillion barrels of oil. We'll use the next trillion in 30. . . . [The] era of easy oil is over.99

—David J. O'Reilly, chairman, Chevron Corporation.

It is difficult for most people to imagine a world without electric lights, automobiles, jet airplanes, and computers. Yet these hallmarks of modern civilization would not run without a constant supply of nonrenewable fossil fuels—oil, natural gas, and coal. Without these energy sources there would be no electricity to power home appliances, no petroleum to fuel trucks, tractors, trains, ships, and cars, and no way to produce the billions of pounds of food people eat every day.

Fossil fuels provide the power that drives contemporary society, and most scientists believe that oil and natural gas supplies will run short within the foreseeable future. There is an ongoing debate as to how long these vital fuel supplies will last and whether alternatives such as solar power or hydrogen fuel cells will be able to take their place. But few doubt that new sources of energy will need to be developed—and old sources will need to be used more efficiently—in order to meet humanity's growing energy demands.

The Power of Petroleum

The dark, slimy liquid known as oil, petroleum, or crude is at the center of the debate over energy and the future. It is produced in nearly every region on earth, including the Americas, Europe, Africa, and Asia; how-

Iraqi workers operate the valves at the oil fields of Rumaila. About 80 percent of the world's known petroleum reserves are in the Middle East.

ever, about 80 percent of the world's known petroleum reserves are in the Middle East. Saudi Arabia alone has one-quarter of the world's oil while Iran, Iraq, the United Arab Emirates, Qatar, and Kuwait are also major producers.

No matter where it is found, oil is a nonrenewable resource that is being consumed at an unprecedented rate. Experts believe that about half of all the recoverable oil on earth has been used since the late 19th century. The consumption of this liquid energy has produced industrial societies that are now completely dependent on a steady supply of petroleum.

Gasoline fuels the 800 million cars, SUVs, and light trucks in use today throughout the world. Diesel fuel is used in agricultural machinery, tractor-trailer trucks, buses, trains, and heating equipment. Kerosene-type jet fuel keeps airplanes flying, while heavy fuel oils are used to power ships. Countless miles of roadways are paved with oil-based asphalt, and refined lubricants keep millions of machines running smoothly.

Petrochemicals, or chemicals made from petroleum, are used to manufacture animal feeds, clothing, building materials, plastics, and even medical products such as antihistamines, antiseptics, and artificial hearts. The average home is also filled with oil-based plastic products, including electronics, kitchenware, food packaging, bedding, furniture, sports equipment, hygiene products, and more. Finally, about one in ten barrels of oil is used to produce electricity in power plants. As energy expert Peter Tertzakian writes in *A Thousand Barrels a Second,* "Every time we flick on a light switch, turn up the heat, or start up our car, a vast and complex energy supply chain kicks into gear."[1]

> "Every time we flick on a light switch, turn up the heat, or start up our car, a vast and complex energy supply chain kicks into gear.

Several factors have contributed to widespread reliance on petroleum. Since the first oil wells went into mass production in 1880, oil has been available in abundance. As a liquid, oil is easy to pump from the ground, store in tanks, and transport via pipelines or tankers to areas where it is needed. Oil also has an extremely high energy density. This means that it is the lightest and most compact substance available that can provide the power found in a tank of gasoline. This is explained by geologist Walter Youngquist, who compares gasoline to standard batteries:

> [A] gallon of gasoline weighing about 8 pounds has the same energy as one ton of conventional lead-acid storage batteries. . . . Even if much improved storage batteries were devised, they cannot compete with gasoline or diesel fuel in energy density. Also . . . there is no battery pack which can effectively move heavy farm machinery over miles of farm fields, and no electric battery system seems even remotely able to propel a Boeing 747 14 hours nonstop at 600 miles an hour from New York to Cape Town [South Africa].[2]

"It's Not Sustainable"

Because of its high energy density and ease of use, petroleum is incredibly versatile. From rotary lawn mowers to 300,000-ton supertanker ships, oil produces power for a wide variety of machines. And the positive aspects of petroleum have created a demand that grows every year. In 1970 worldwide oil consumption was about 43 million barrels a day. In 2005 people were consuming nearly twice as much oil, 84 million barrels a day. And the annual demand is expected to increase by 2 million barrels a day for at least the next 20 years. This means that by 2025 people will burn 40 percent more oil than was used in 2005.

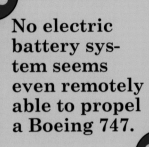

No electric battery system seems even remotely able to propel a Boeing 747.

While demand for petroleum is growing, the most productive oil fields are producing less oil every year. Oil production in the United States and Canada peaked in 1970 and 1973, respectively, and has been declining ever since. Great Britain's oil production in the North Sea has been falling since the late 1990s, while two of China's largest producing oil fields are also in decline. Mexico, which is the third largest oil supplier to the United States after Canada, has discovered that its Cantarell oil field peaked in late 2004. Production at Cantarell, the world's second largest oil complex, is now declining by 14 percent a year.

Because of such declines other oil producers will need to increase production by 6 to 8 million barrels of oil per day if patterns of depletion and demand remain unchanged. These figures are put into perspective by Sadad al-Husseini, an executive at Aramco, the national oil company of Saudi Arabia, which produces 10 million barrels a day: "That's like a whole new Saudi Arabia every couple of years. It can't be done indefinitely. It's not sustainable."[3]

Clean and Efficient Natural Gas

After petroleum, natural gas is the second most important energy source in the world. Every day, consumers burn about 240 billion cubic feet of natural gas. As with petroleum, the debate over natural gas concerns shrinking supplies and a demand that has been growing at 2 percent annually for the last 20 years.

Natural gas is odorless, colorless, and tasteless and consists primarily of methane, extracted from the earth, like oil. (The smell of rotten eggs associated with natural gas is a chemical odorant added as a safety measure so leaking gas can be detected.) As an energy source, natural gas is the least polluting of the fossil fuels. For example, when compared with coal, natural gas produces much lower levels of the toxic air pollutant sulfur dioxide and half as much of the global warming gas carbon dioxide.

As a cleaner-burning fuel, natural gas can be used indoors. In the United States and Canada, more than half of all homes use natural gas to provide heat, dry clothes, and cook food. About 70 percent of all new homes built are equipped for natural gas use.

Power plants are also large consumers of natural gas because the fuel produces electricity more efficiently than coal or oil. For this reason, natural gas is used to produce about 40 percent of all electricity worldwide. In the United States one in five electrical generating plants relies on natural gas as a fuel source. And nearly all new electric plants being built in the United States are fueled with natural gas.

Besides generating electricity, natural gas has many important industrial uses. In the United States nearly half of all natural gas is used to manufacture pulp and paper, cement, chemicals, plastics, and refined petroleum products such as lubricants. Natural gas also plays a key role in the worldwide production of agricultural fertilizers, as environmental researcher Julian Darley explains in *High Noon for Natural Gas:*

> [The] growth of the human population throughout the twentieth century into the twenty-first is due in great part to the use of industrial fertilizer. One of the most vital, and presently irreplaceable, feedstocks of that fertilizer is natural gas, which makes natural gas possibly the single most critical ingredient in the diet of many human beings.[4]

The Natural Gas Peak

Like oil, natural gas is a vital resource that is in decline. In Great Britain, North Sea natural gas reserves discovered in the 1960s are expected to be gone by 2035. In the rest of Europe, only Norway is expected to have enough domestic gas by 2010. In the United States, oil giant Exxon's former chief executive Lee Raymond said in June 2005 that natural gas supplies have already peaked even as demand continues to grow by more than 2 percent a year.

The loss of locally produced natural gas is a problem because in its original form, gas can only be moved by pipelines. For example, in the United States, millions of miles of pipelines transport natural gas from gas wells to consumers across the country. To move natural gas without pipelines, say from Saudi Arabia to the United States, the

> " Exxon's former chief executive, Lee Raymond said in June 2005 that natural gas supplies have already peaked even as demand continues to grow. "

gas needs to be cooled to very cold temperatures, about minus 260 degrees Fahrenheit (-162°C). This expensive, energy-consuming process turns the gas into liquefied natural gas (LNG). Specially equipped ships and ports are needed to transport and offload LNG, and a single LNG terminal with tankers, pipelines, and processing facilities costs from $8 billion to $10 billion. LNG terminals also take years to build, and since they emit large amounts of air pollution they are almost always opposed by local communities. Because of the problems associated with LNG terminals, less than 5 percent of the natural gas sold worldwide is handled in this manner.

No Shortage of Coal

Coal, the world's third largest source of energy, is the only fossil fuel that is not experiencing a production decline. Across the globe 14 million tons of the hard, black rock is burned every day. While some is used for heating, coal is mainly burned to generate electricity. In China two out of three power plants use coal, while in the United States it is burned by half of all power plants. The importance of this energy source is explained by journalist Jeff Goodell in *Big Coal*:

> More than one hundred years after Thomas Edison connected the first light bulb to a coal-fired generator, coal

remains the bedrock of the electric power industry in America. . . . [We] burn more than a billion tons of it a year.[5]

As with oil and natural gas, there are good reasons why coal plays a central role in modern life. Besides being easy to transport, store, and burn, coal is the most plentiful traditional energy source. An estimated 1 trillion tons of coal is buried in the earth and this abundance makes it the cheapest fossil fuel. As Goodell writes: "In a world starved for energy, the importance of this simple fact cannot be underestimated: the world needs cheap power, and coal can provide it."[6]

> **More than 100 years after Thomas Edison connected the first light bulb to a coal-fired generator, coal remains the bedrock of the electric power industry in America.**

The United States has some of the largest coal reserves—about 25 percent of the world total, or 270 billion tons. This is enough, at the current rate of use, to supply the United States for another 250 years. Other nations with large reserves include Russia with 176 billion tons and China with 126 billion tons. By comparison, Western Europe only has about 36 billion tons of coal.

Pollution Problems

Because of its abundance, scientists see coal as a fuel that may someday alleviate a possible oil or natural gas crisis. As a replacement for oil, coal can be liquefied through a complicated industrial process. In order to be used in place of natural gas, coal can be converted to gas by subjecting it to pressure and high temperatures; however, converting coal to gas or liquid is expensive, and oil prices would have to climb drastically and remain high for such processes to make economic sense. In addition, coal has a much lower energy density than oil, and producing liquefied coal in sufficient quantities to replace gasoline would require mining projects on a scale never yet attempted.

Even at its current levels, coal mining is extremely damaging to the environment. Since coal is buried underground, the most efficient method to mine it is through a process called mountaintop removal (MTR). This process, practiced in Kentucky and West Virginia, utilizes industrial explosives to blow off the top of a mountain. The resulting rubble is pushed into adjacent valleys, causing complete ecological destruction.

Mountaintop removal is a localized problem, but burning coal has another significant drawback. Coal is dirty and emits high concentrations of carbon dioxide when burned. Burning coal also produces sulfur dioxide, which turns into sulfuric acid when mixed with water vapor in the air. This falls to earth as acid rain, which kills trees and plants, pollutes waterways, and damages buildings and monuments.

The Role of Nuclear Energy

Pollution problems and fears of shortages have left experts searching for alternatives to coal, natural gas, and oil. Some believe that nuclear power has the greatest potential because 2 pounds of uranium has 300 million times more energy density than a similar amount of coal.

Nuclear power plants produce about 16 percent of the world's electricity by utilizing the energy in uranium to boil water. The resulting steam drives turbines to produce electricity. Although uranium is long lasting, after 3 to 6 years it is no longer useful for electrical generation and must be replaced. As such, 25 to 33 percent of the uranium used in a nuclear reactor has to be replaced annually. With 441 commercial reactors operating in 31 countries more than 472,000 pounds of uranium is consumed every day.

The spent fuel rods from nuclear reactors continue to generate intense heat for centuries and dangerous radiation for at least 30,000 years. While most nuclear proponents advocate burying the waste, there are fears that repositories can leak and poison underground water deposits. This fact has created a political backlash against nuclear power in the United States and Europe; however, Japan and China are developing a new

> " With 441 commercial reactors operating in 31 countries more than 472,000 pounds of uranium is consumed every day. "

generation of nuclear power plants expected to be safer and more efficient than those operating today.

South Korea, whose demand for electricity is increasing by nearly 10 percent a year, is a good example of a nation that is investing in nuclear power. In 2006 South Korea's 9 nuclear power plants produced 40 percent of the nation's electricity. In addition, 9 more plants were under construction and expected to go online by 2015. For a nation that has to import 97 percent of its coal and oil, these plants have allowed the South Korean economy to prosper. Nuclear waste is being stored at individual plants but there are plans to build several underground waste disposal repositories.

Hydrogen Power

The problems associated with nuclear power have led some to believe that it is only a temporary solution to a possible long-term energy crisis. Thus, there is a widespread movement to use environmentally friendly, sustainable sources of power such as sunlight, water, and wind. Proponents say these energy sources could reduce the widespread reliance on fossil fuels and prevent disaster when oil supplies run low.

> There is a widespread movement to use environmentally friendly, sustainable sources of power such as sunlight, water, and wind.

Because society is so dependent on cars, trucks, and buses, several major auto companies are working to replace gasoline engines with fuel cells. These high-tech batteries run on hydrogen gas molecules obtained from water. While this technology is promising, the infrastructure is not in place to produce, ship, and store highly flammable hydrogen. Advocates of fuel cells are hoping to put the infrastructure in place in the coming decades before a potential gas shortage disrupts the world economy.

Renewable Energy

Utilizing renewable solar power from the sun is seen as another way to reduce dependence on fossil fuels and nuclear energy. Every day the sun provides more energy than the entire population of the earth could consume in 27 years. Solar power can be transformed into electricity through the use of photovoltaic (PV) cells, and the average home has more than enough roof space for a solar cell that would supply most of its power needs. At current electricity prices, however, these expensive cells are not competitive with traditional sources of electricity. In addition, solar energy can only be used during the daytime as there is no economical way to store the electrical energy for use at night, on cloudy days, or during snowstorms. Therefore, if millions more people used PV cells, traditional power plants would still be needed.

Some believe that modern windmills, called wind turbines, are less problematic than solar cells. These machines use large propellers to convert wind into electricity. While they are generally efficient, wind turbines have been a source of contention in some communities. The propellers on the 40-story wind machines can be extremely loud. Because they are often located in windy passes that are used by birds and bats as flyways, wind turbines kill thousands of flying creatures every year.

Although there are problems associated with wind turbines and other renewable energy sources, advocates believe that a move to renewables is the only way to prevent an energy crisis. Commenting on this transition, Youngquist writes:

> A realistic appraisal of the future encourages people to properly prepare for the coming events. Delay in dealing with the issues will surely result in unpleasant surprises. Let us get on with the task of moving orderly into the post-petroleum paradigm.[7]

Is the World Running Out of Oil?

66 We've embarked on the beginning of the last days of the age of oil.99

—Mike Bowlin, chairman and CEO, ARCO.

On any given day, thousands of oil geologists and geophysicists who work for major energy companies are combing the earth in search of new oil supplies. With the help of satellites, sophisticated sounding techniques, and advanced computer technology, the scientists create 3-dimensional, virtual-reality maps of underground oil reservoirs. These maps help oil company executives decide where to spend tens of millions of dollars drilling for oil.

While high-tech methods are helping to provide new sources of petroleum, they have also produced startling information concerning existing oil supplies. By utilizing advanced exploration technology, analysts have arrived at the conclusion that 90 percent of the world's oil resources have already been discovered. Energy expert Peter Tertzakian explains what this means for the future:

> Talk to any petroleum geologist or geophysicist today and you will hear the same thing. Nearly all the really big "elephant" oil fields, the ones that contain billions of barrels of reserves, have been identified. . . . [Today's new] oil fields are increasingly smaller in size. A new oil field containing a few hundred million barrels of reserves is big news. At the current rate of global consumption, such fields would be drained in days if we could turn on a spigot.[8]

Light Sweet Crude

When geologists speak of elephant fields they are talking about huge reservoirs of high-grade oil called light sweet crude. This type of oil has been supplying power needs for the last 100 years. It has a low sulfur content and is easily refined into gasoline. The entire oil infrastructure, built to deliver billions of gallons of gas to motorists every day, has been designed to process light sweet crude.

In addition to its use for transportation, light sweet crude is central to the production of food. With the global population expected to nearly double from around 6 billion today to 11 billion in 2040, oil production will need to increase 1000 percent from current levels just to feed the earth's burgeoning populace.

Few analysts believe this is possible. Extracting oil from the ground is an extremely difficult and costly process. Oil reservoirs do not resemble huge underground lakes filled with petroleum. Instead, they are similar to sponges with thousands of holes filled with varying amounts of liquid. When a well is drilled into an underground cavity, pressure within the earth will usually force the oil to the surface to create a gusher. After several weeks, months, or years, however, the natural pressure dissipates and secondary recovery efforts must begin. This involves expensive and energy-intensive methods like pumping water or natural gas into the reservoir to increase underground pressure. This process is often volatile and erratic. If oil is extracted too quickly or too much pressure is placed on the reservoir, it can collapse or contaminate the oil with mud, sand, or liquids.

> " **Nearly all the really big 'elephant' oil fields, the ones that contain billions of barrels of reserves, have been identified.** "

Because of extraction problems, only about 35 percent of petroleum in any given reservoir can be recovered, leaving nearly two-thirds of the world's oil unrecoverable underground. So, while there will always be oil in the earth, it may not be economically—or technologically—feasible to pump it out.

Peak Oil Theory

A group of people known as peak oilers think that oil supplies will fall short in a matter of a few decades as global demand surges. Peak oilers base their claims on the work of Marion King Hubbert, a Shell Oil geologist. In 1956 Hubbert formulated a complex mathematical theory, known as Hubbert's Peak, that predicted that U.S. oil production would peak in the 1970s. Hubbert's Peak was rejected by almost everyone in the oil industry. In 1973, however, U.S. production of crude oil indeed began to fall and has been in decline ever since.

> "A group of people known as peak oilers think that oil supplies will fall short in a matter of a few decades as global demand surges."

Researchers have applied Hubbert's formula to worldwide petroleum production. They have suggested that global production could peak sometime between 2008 and 2020. Even the U.S. Department of Energy is forecasting a world oil peak in 2037.

Troubles in Saudi Arabia

Those who believe in the peak oil theory say that problems will begin with Saudi Arabia, a nation that produced 10 percent of the world's daily supply of oil in 2006. According to the U.S. Department of Energy (DOE), the Saudis will need to double production to meet global demand by 2020; however, some experts warn that the Saudis will not be able to do so because all of their elephant oil fields are 40 to 60 years old. Although they are not currently in decline, some predict Saudi production will peak within 10 years. This would spark an unprecedented worldwide energy crisis, described by journalist Peter Maass in the *New York Times*:

> [The] price of a barrel of oil could soar to triple-digit levels. This, in turn, could bring on a global recession, a result of exorbitant prices for transport fuels and for products that rely on petrochemicals—which is to say, almost every product on the market. The impact on the

American way of life would be profound: cars cannot be propelled by roof-borne windmills. The suburban and exurban lifestyles, hinged to two-car families and constant trips to work, school and Wal-Mart, might become unaffordable or, if gas rationing is imposed, impossible. . . . [The] cost of home heating would soar.[9]

A More Optimistic View

Those who have a more optimistic view point out that there is an enormous supply of oil left underground, enough to meet world energy needs for at least another 30 years. This oil could last several decades longer if conservation measures, such as heavily taxing gasoline or improving automobile gas mileage, were put in place. In addition, new drilling technology is also expected to allow oil companies to recover at least some of the 65 percent of oil left behind by present extraction techniques. Geophysicist David Deming explains: "Every year, technological advances make it possible to draw upon petroleum resources whose extraction was once unthinkable."[10]

Oil companies now commonly drill wells 30,000 feet deep, a 30 percent increase since the early 1990s. And new drilling records continue to be set. In December 2005, Chevron sank a well 34,189 feet deep in the Gulf of Mexico. Little more than 9 months later, Chevron discovered a major oil field in the gulf, about 270 miles southwest of New Orleans. The deposit, nicknamed Jack, is under 7,000 feet of water and more than 20,000 feet below the sea floor. Experts believe that Jack contains 3 to 15 billion barrels of oil. In the coming years, the oil field could produce 750,000 barrels of oil a day, about 3.5 percent of America's daily needs.

> " Every year, technological advances make it possible to draw upon petroleum resources whose extraction was once unthinkable. "

Other Sources of Oil

Even with new oil discoveries like Jack, the supply of light sweet crude is slowly dwindling. For this reason, energy companies are working with nontraditional sources that were once considered too expensive. For example, in recent years, gooey black sands, called oil sands, tar sands, or bituminous sands, have set off an energy boom in Alberta, Canada.

Oil sands contain clay, sand, water, and bitumen, a semisolid form of degraded oil. Bitumen cannot be recovered through drilling and must be mined. It is then processed into a synthetic crude oil, or syncrude, that can be converted at special refineries into gasoline, diesel, and other petroleum products. Because of the logistics of mining and refining bitumen, it is an unprofitable venture when oil prices are low; however, the increase in oil prices between 2000 and 2006 made the production of syncrude an attractive proposition for energy companies.

Few doubt that Canadian oil sands can help offset an energy crisis. According to some estimates the sands contain as much as one-third of all the oil on earth. At current levels, Canadian companies are only producing about 1 percent of world consumption; however, production is expected to quadruple to about 4 percent by 2015.

Extraheavy Oil

Rising oil prices are also fueling Venezuela's move to utilize its reserves of oil sand and its supply of extraheavy crude oil. One operation uses oil sands to create a new product called Orimulsion. This substance is a mixture of 70 percent bitumen, 29 percent water, and 1 percent emulsifiers that enable the oil and water to mix. Orimulsion is made to be used as fuel oil in power plants. Other nations, such as China, Denmark, Italy, and Japan, are working to formulate their own types of Orimulsion for industrial use.

Venezuela is also finding ways to refine normally unusable extraheavy crude by mixing it with light crude and the petroleum-based solvent naphtha. When mixed with these substances, the extraheavy oil can be transported through regular pipelines to refineries where it is made into syncrude. By using new technologies and formulating Orimulsion, it is estimated that Venezuela can produce nearly as much oil as the Canadian bitumen sands in the coming century. Together, these new sources could supply more than 8 percent of the world's current oil demands, nearly as much as Saudi Arabia does now.

Environmental Problems

Oil sands are recovered through extremely destructive strip-mining techniques. In Alberta, thousands of acres of forests, bogs, and rivers have been destroyed by the oil boom. Furthermore, processing oil sands takes an incredible amount of energy. As former vice president Al Gore told *Rolling Stone* in July 2006: "For every barrel of oil they extract there, they have to use enough natural gas to heat a family's home for four days. And they have to tear up four tons of landscape, all for one barrel of oil. It is truly nuts."[11] This process also causes a great deal of air pollution. For each barrel of oil produced, more than 175 pounds of CO_2 is released into the atmosphere. In addition, oil sands production is extremely water intensive: About 4 barrels of wastewater are produced for every barrel of oil.

> **For every barrel of oil they extract [from oil sands], they have to use enough natural gas to heat a family's home for four days.**

China's Growing Thirst for Oil

Whatever the problems with oil sands, there is little doubt that the oil found in Alberta and Venezuela will be in great demand in the coming years. Figures show that consumers in 2006 burned more than four barrels of oil for every new barrel found. Since the 1990s, new oil discoveries have been limited to about 10 billion barrels a year, about one-sixth of the amount discovered annually in the 1950s and 1960s. Meanwhile, production in most major fields is declining by an average of 4 to 5 percent a year as global oil demand increases at 2 to 3 percent annually.

Much of the demand for more oil is coming from China, a nation where few people owned cars until recently. In 2004, however, the Chinese were buying 400,000 cars a month, and the nation surpassed Japan as the world's second largest consumer of petroleum. By 2006 China was consuming 6.5 million barrels of oil per day. While this is about one-third of the amount of oil used by the United States every day, it is a dramatic increase from the 2 million barrels China burned daily in 1990. And experts predict that China's consumption will more than double by

2025 as the country becomes one of the largest producers of manufactured goods on earth.

China's growth as a major oil consumer was not foreseen by energy experts as little as ten years ago. But like many aspects of the oil business, issues of supply and demand are unpredictable, and experts say that the oil market will remain volatile for many years to come. A world energy crisis could be triggered by a number of unforeseeable factors that might include a terrorist attack on Saudi oil facilities, a war in Iran, or a major hurricane in the Gulf of Mexico. Even without such disasters, there is little doubt that oil reserves are shrinking every day. Whether or not industrialized society will adjust to these changing conditions remains to be seen.

Is the World Running Out of Oil?

> **When the global peak in oil production is reached . . . it will be difficult or impossible to pump as much as the year before.**

—Richard Heinberg, "The Party's Over," Ecomall, 2006. www.ecomall.com.

Heinberg is an author and energy resource expert.

> **The only thing the public is going to remember is that peak oil was just like Y2K—a bunch of squealing Chicken Littles whose predictions turned out wrong.**

—JD, "Peak Oil Is Dead," blogspot.com, September 8, 2005. http://peakoildebunked.blogspot.com.

JD is a blogger and creator of the Peak Oil Debunked Web site.

> **America is addicted to oil.**

—George W. Bush, "State of the Union Address by the President," January 31, 2006. www.whitehouse.gov.

Bush is the 43rd president of the United States.

* Editor's Note: While the definition of a primary source can be narrowly or broadly defined, for the purposes of Compact Research, a primary source consists of: 1) results of original research presented by an organization or researcher; 2) eyewitness accounts of events, personal experience, or work experience; 3) first-person editorials offering pundits' opinions; 4) government officials presenting political plans and/or policies; 5) representatives of organizations presenting testimony or policy.

❝One fact is undeniable: over the past decade, oil production has been falling in 33 of the world's 48 largest oil producing countries.❞

—Christopher Flavin, "Over the Peak," *World Watch*, January/February 2006, p. 1.

Flavin is president of the Worldwatch Institute, a research institute concerned with environmental, social, and economic trends.

....

❝From the moment we wake up in the morning to the moment we go to sleep, oil controls our lives.❞

—Matthew Yeomans, *Oil: Anatomy of an Industry.* New York: New Press, 2004, p. xi.

Yeomans is a journalist who specializes in the politics of oil.

....

❝The United States would be all but powerless to protect the American economy in the face of a catastrophic disruption of oil markets.❞

—John Mintz, "Outcome Grim at Oil War Game: Former Officials Fail to Prevent Recession in Mock Energy Crisis," *Washington Post,* June 24, 2005, p. A19.

Mintz is a *Washington Post* staff writer.

....

❝As time goes on, energy needs will grow. Eventually, the resources we use now will prove insufficient, or will [be too costly].❞

—Lee R. Raymond, "Challenges in Meeting Future Global Energy Demand," ExxonMobil.com, November 4, 2004. http://exxonmobil.com.

Raymond is the former CEO of ExxonMobil.

....

❝We are confident there is more oil to be found in Saudi Arabia. There are vast areas of Saudi Arabia yet to be explored. They present great opportunities for new discoveries.❞

—Ali al-Naimi, "U.S.-Saudi Relations and Global Energy Security," Saudi-US Relations Information Service, May 7, 2004. www.saudi-us-relations.org.

Al-Naimi is the Saudi oil minister.

❝The entire world assumes Saudi Arabia can carry everyone's energy needs on its back cheaply. If this turns out to not work, there is no Plan B, and the world will face a giant energy crisis.❞

—Matthew Simmons, "Saudi's Missing Barrels of Oil Production," copvcia.com, 2004. www.copvcia.com.

Simmons is an energy industry analyst.

❝We need to do more to encourage—and bring to the mainstream—conservation. Certainly in the near-term, conservation is the easiest, cheapest and most reliable 'new' energy source there is.❞

—David J. O'Reilly, "Global Energy: The New Equation," Chevron, July 23, 2004. www.chevron.com.

O'Reilly is chairman and CEO of Chevron Texaco Corporation.

❝This is not the first time that the world has 'run out of oil.' It's more like the fifth. Cycles of shortage and surplus characterize the entire history of the oil industry.❞

—Daniel Yergin, "It's Not the End of the Oil Age," *Washington Post*, July 31, 2005, p. B7.

Yergin is the author of the Pulitzer prize–winning book *The Prize: The Epic Quest for Oil, Money, and Power.*

Facts and Illustrations

Is the World Running Out of Oil?

- In 2006 the United States consumed 25 percent of the world's oil.

- The world consumes more than 26 billion barrels of oil a year, but new discoveries average only 7 billion barrels a year.

- According to the U.S. Department of Energy, to meet growing demands, total world petroleum output will have to grow by 44 million barrels a day between now and 2025.

- To increase output by 44 million barrels a day, 4 reservoirs the size of those in Saudi Arabia would need to be found.

- Ninety percent of the world's oil reserves have already been discovered.

- New oil discoveries have been limited to about 10 billion barrels a year, compared with 60 billion barrels annually in the 1950s and 1960s.

- The U.S. Department of Energy predicts that global oil production will peak in 2037.

- Between September 2003 and July 2006 the price of oil more than tripled, jumping from $24 a barrel to $78.40.

World Oil and Gas Production, 1930–2050

This graph shows a projection for world oil and gas production through 2050. It suggests production is currently reaching a peak and will begin to decline shortly before 2010. With current technology, it is difficult to determine world oil supply and reserves. Other experts project that the peak will not come for several decades.

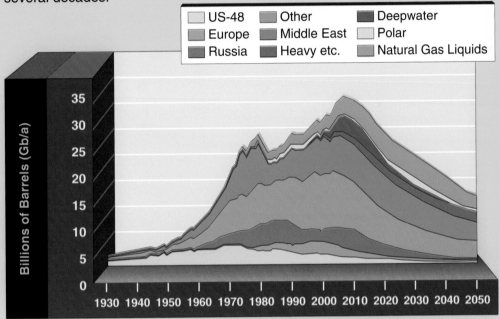

Source: Colin J. Campbell, "The Heart of the Matter," The Association for the Study of Peak Oil and Gas, 2003.

U.S. Oil Production Decreases While Use Increases

Oil production in the United States peaked in 1970 and has been declining ever since. If patterns of depletion and demand remain unchanged, producers will have to increase production by 6 to 8 million barrels per day.

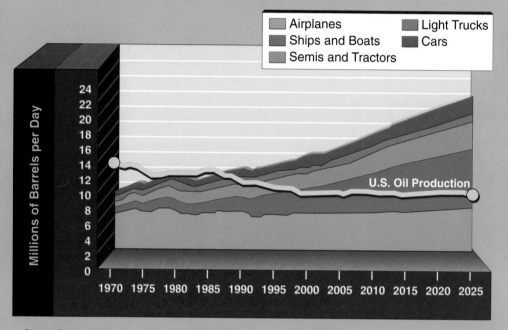

Source: Center for Transportation Analysis, *Transportation Energy Data Book,* edition 22, September 2002, and Energy Information Administration, *EIA Annual Energy Outlook, 2003,* January 2003.

- The record for the world's deepest oil well is 34,189 feet, drilled in the Gulf of Mexico for Chevron in December 2005.

- The Ghawar oil field in Saudi Arabia is the largest and most productive oil field in the world.

Saudi Arabia Has the Most Oil Reserves

Aramco, the national oil company of Saudi Arabia, produces 10 million barrels of oil each day.

COUNTRY	OIL RESERVES "Billions of Barrels"
Saudi Arabia	264.3
Canada	178.8
Iran	132.5
Iraq	115.0
Kuwait	101.5
UAE	97.8
Venezuela	79.7
Russia	60.0
Libya	39.1
Nigeria	35.9
United States	21.4
China	18.3
Qatar	15.2
Mexico	12.9
Algeria	11.4
Brazil	11.2
Kazakhstan	9.0
Norway	7.7
Azerbaijan	7.0
India	5.8
Rest of World	68.1
World Total	**1,292.6**

Source: *Oil and Gas Journal*, "Worldwide Look at Reserves and Production," vol. 103, no. 47, December 19, 2005, pp. 24–25.

Top 20 Oil Producers and Consumers

In 2002, the U.S. was the largest producer of oil in the world, followed by Saudi Arabia and Russia; however, the U.S. consumes more than twice the oil it produces. Experts believe that about half of all the available oil worldwide has been used since the late 19th century.

Top 20 Producers

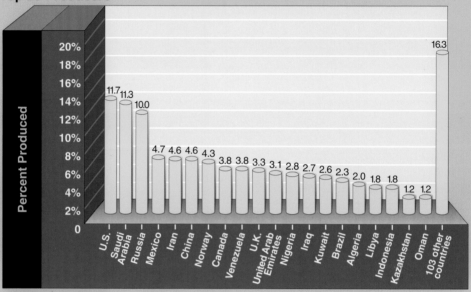

Top 20 Consumers

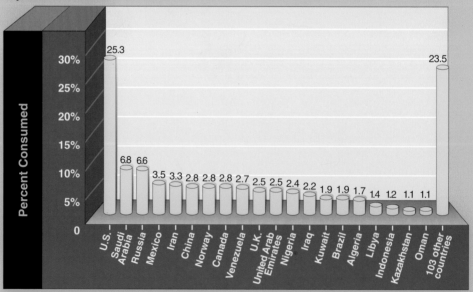

Source: Robert L. Hirsch, "Peaking of World Oil Production: Impacts, Mitigation, and Risk Management," February 2005. www.oilcrash.com.

Will Alternative-Fuel Vehicles Solve the World Energy Crisis?

> 66 There is fuel in every bit of vegetable matter that can be fermented. There's enough alcohol in one year's yield of an acre of potatoes to drive the machinery necessary to cultivate the fields for a hundred years. 99
>
> —Henry Ford, founder of the Ford Motor Company, 1925.

Between September 2003 and July 2006 the average price of gas in the United States jumped from about $1.45 a gallon to over $3 a gallon. During that same period, sales of hybrid cars, which combine power from gasoline engines and electric batteries to deliver more than 50 miles per gallon (mpg), quadrupled. By early 2006 more than 212,000 hybrid vehicles were traveling America's roads. Manufacturers could not make the cars fast enough, and in California there was a 6-month waiting list for the Toyota Prius hybrid. As the fastest-growing segment of the new-car market, experts predict sales of hybrids will triple in the United States before 2012.

Hybrids are also growing in popularity in other parts of the world. For example, in Germany, where gas is close to $7 a gallon, about 70 percent of all new cars sold by 2010 are expected to be hybrids. And some analysts predict that all private vehicles will be hybrids by 2025. If this is accurate, cars of the future will get two to three times the gas mileage of cars today, which could slow or prevent a world gas crisis in the coming decades.

A Variety of Fuels

Hybrids are the latest in a long line of cars that have been produced to reduce dependence on oil. And throughout automotive history, there have

> As the fastest-growing segment of the new-car market, experts predict sales of hybrids will triple in the United States before 2012.

been many cars that did not require gasoline at all. When automobiles were first introduced in the late 1800s, motorists could choose cars that were powered by electric batteries, steam, and even peanut oil (now called biofuel). However, when Oldsmobile began mass-producing gasoline-powered cars in 1902, and Henry Ford set up his Model T Ford assembly line in 1908, alternative-fuel vehicles were largely forgotten. Although Ford originally wanted the Model T to run on ethanol (grain alcohol), major oil discoveries in Texas in the early 1900s made gasoline widely available and inexpensive. It would be decades before anyone would propose running a motor vehicle on any fuel besides petroleum.

In the mid-1970s people once again began to consider alternative-fuel vehicles due to environmental concerns and fears of oil shortages. Since that time a host of propulsion systems have been employed by inventors, researchers, and even major auto makers such as General Motors, Toyota, Ford, and Honda. In the 21st century, motorists can choose from cars that are powered by batteries, alcohol, biofuels, and hydrogen. And in 2006 millions of flexible-fuel vehicles, engineered to run on gasoline, alcohol, or a combination of the two are on the road.

All of the alternative-fuel vehicles have their benefits and drawbacks, and some are only in experimental stages; however, few doubt that alternative-fuel vehicles will be necessary in the future when gas prices increase as oil supplies diminish.

Alcohol as a Fuel

Currently the most widely used alternative fuel is ethanol. Also known as ethyl alcohol, ethanol is the same substance found in alcoholic beverages and is distilled from grains, sugar cane, and other vegetable matter.

Ethanol is used as fuel in at least 30 countries throughout the world, including India, Sweden, France, Canada, China, Colombia, and the United States. In 2006 Brazil was the world's largest producer of fuel ethanol, making 3.6 billion gallons annually from sugar cane. Brazil began switching from gasoline to ethanol in the mid-1970s during a global oil shortage. After more than 30 years, the nation achieved "energy independence," or freedom from reliance on imported oil, in 2006.

Brazilian motorists who drive older cars use a fuel called E25 that is 25 percent ethanol and 75 percent gasoline. Cars built since 2003 are flexible-fuel vehicles that can run on E25 or E100, 100 percent ethanol.

Ethanol from Corn

In the United States, corn is used to produce ethanol, and distillers make about 4.5 billion gallons annually. While this is more ethanol than is produced in Brazil, the United States economy is 10 times bigger and Americans own more than 7 times more cars. Therefore, American ethanol production only meets about 3 percent of the nation's transportation needs. If the United States tried to mimic Brazil's ethanol success, it would require 5 times more corn than is now produced in America every year. Commenting on this problem, Ronald Bailey, science correspondent for *Reason* magazine, writes: "This would also leave no corn for food . . . [and burning] food for fuel raises some interesting moral questions in a world in which 800 million people are still malnourished."[12]

Scientists are trying to invent new and better ways to produce ethanol. One process would make use of waste products, such as plant stems and leaves. Other ethanol production methods would use the energy found in nonfood items. For example, in his 2006 State of the Union address, President George W. Bush mentioned that a common prairie grass called switch grass could produce enough ethanol to help end America's reliance on imported oil. However, Bush's plan has severe limitations: It would take 100 million acres of switch grass—an area the size of California—to produce only 20 percent of the ethanol Americans currently use as a fuel additive.

> **[Burning] food for fuel raises some interesting moral questions in a world in which 800 million people are still malnourished.**

Problems with Ethanol

Land use issues are only part of the ethanol problem. Critics also point out that ethanol is uneconomical when used as a gas substitute. In October 2006, *Consumer Reports* ran a wide variety of tests on E85 cars and discovered that they get much lower fuel economy than vehicles using all-gasoline fuel. For example, a Chevrolet Tahoe FFV got only 10 mpg on E85 while a traditional Tahoe achieves 14 mpg with gasoline.

Manufacturing ethanol is also an extremely inefficient process and requires large amounts of oil, natural gas, fertilizers, and other fossil fuels. This point is made by two of ethanol's most vocal opponents, Tad Patzek, an engineering professor at the University of California–Berkeley and David Pimentel, a professor of ecology at Cornell University. In 2005 Patzek and Pimentel studied every aspect of ethanol crop production, including planting, fertilizing, irrigating, harvesting, grinding, and distilling. They discovered that corn ethanol requires 29 percent more fossil fuel than the energy contained in the ethanol. The professors also studied other crops, such as soybeans and sunflower seeds, and came to similar conclusions. When discussing his findings, Patzek says that if the United States channeled the billions spent to subsidize ethanol production into fuel-efficient cars and solar cells, "that would give us so much more bang for the buck that it's a no-brainer."[13]

Not everyone agrees with the findings of Pimentel and Patzek. Proponents of ethanol claim that the process yields 25 percent more energy than it consumes; however, the fact remains that if there was no oil, there could be no ethanol. Whatever the case, General Motors announced in 2005 that their entire fleet would be E85-ready by 2010 and there is little doubt that alcohol will be filling millions of gas tanks in the coming years.

Biodiesel

Vegetable matter can also be used to make fuel. And amazingly, modern diesel engines can run on 100 percent vegetable- or animal-based fuel called biodiesel. This substance, called B100, is made from cooking oil that has been processed to remove the thick, sticky glycerin. B100 fuel can be created from canola, soybean, hemp, palm, and algae oils along with animal fats such as tallow, lard, and fish oil.

Biodiesel is much more efficient than ethanol to produce because it has a very high energy yield. For example it only takes about 1 gallon of gas to produce 3 gallons of biodiesel. While most cars in North America do not have diesel engines, biodiesel can be used in the millions of trucks and buses currently on the road. And in Europe, where diesel fuel is cheaper than gasoline, about half of all cars have diesel engines that could be run on B100.

Oily Algae

Like corn ethanol, biodiesel uses food crops for fuel, and there is not enough farmland in the world to produce large quantities of B100. Research has shown, however, that certain algae species contain over 50 percent oil. Growing algae for biodiesel would not require valuable farmland since the organisms can be grown at sewage treatment plants in wastewater.

Research has also shown that algae could be produced in shallow saltwater pools on algae farms located in desolate desert areas. Using the Sonoran Desert in Arizona and California as an example, scientists estimate that about 12 percent of the desert could someday produce enough biodiesel to replace all of the petroleum used in the United States; however, this would require millions of motorists to buy new cars with diesel engines, a process that could take decades.

> " Growing algae for biodiesel would not require valuable farmland since the organisms can be grown at sewage treatment plants in wastewater. "

Fry-Grease Fuel

Some people are not waiting for the construction of algae farms to run their vehicles on vegetable oil. Instead they run their diesel cars and trucks on straight vegetable oil (SVO) poured directly out of containers purchased at grocery stores. Unlike biodiesel, which is a processed fuel that can be used in any diesel engine, those who use SVO must make some modifications to their vehicles at a cost of about $700 in 2006.

Vehicles that run on SVO can also be fueled with waste vegetable oil (WVO). This substance is simply used fry-grease recycled from food processing plants and restaurants. While only several thousand people have converted their diesel engines to use WVO fuel, those who have believe it is the answer to energy independence. As biodiesel proponent Jules Dervaes writes in the Path to Freedom Web site: "Forget sky-high gas prices, dump the petroleum addiction and switch to clean and renewable biofuels. Biodiesel is a viable, sustainable alternative to petroleum."[14]

Hydrogen Fuel Cells

Despite the optimism exhibited by WVO enthusiasts, all of the waste oil generated by the food industry in the United States would only produce enough fuel to run 1 percent of the nation's cars. Hydrogen, however, a combustible element of water, is one of the most ubiquitous substances on earth. It can be used in 2 ways to fuel cars. In 2007 German automaker BMW is introducing the BMW Hydrogen 7 Saloon. This vehicle uses hydrogen gas in the same way some vehicles burn natural gas—to power a traditional internal combustion engine. Because there is little hydrogen infrastructure in place, however, the Hydrogen 7 Saloon has 2 tanks. Drivers can switch between hydrogen or regular gasoline to avoid being stranded.

Japanese automaker Honda plans to use hydrogen in a different way. In 2008 Honda will begin producing a sleek sports sedan that converts hydrogen to electricity through fuel cells that power electric motors.

Whether used to power fuel cells or burned as a gas, many believe that a switch to hydrogen will prevent an impending petroleum crisis. As environmental author Jeremy Rifkin explains:

> [Hydrogen] never runs out and produces no harmful CO_2 emissions when burned; the only byproducts are heat and pure water. That is why it's been called "the forever fuel."[15]

Like other fuel sources, hydrogen has its detractors who believe that the fuel will not solve future energy problems. Current methods of separating hydrogen molecules from water depend on electricity, which is most often produced with fossil fuels. Wind and solar power might be used to produce hydrogen, but those technologies are not currently available on a large scale. In addition, fuel cells are extremely expensive for use in mainstream applications like cars because they are made with platinum, the same metal used in expensive jewelry. Finally, hydrogen is extremely flammable, generates great heat, and is dangerous to handle.

Promoters of hydrogen, ethanol, and biodiesel understand that there is no single, extremely effective solution to what has been called society's addiction to oil; however, experts believe that new methods need to be developed and put into place as quickly as possible. With energy experts predicting oil shortages in the next thirty years, the development of alternative-fuel vehicles is deemed necessary and essential by everyone from staunch environmentalists to auto industry executives.

> [Hydrogen] never runs out and produces no harmful CO_2 emissions when burned; the only by-products are heat and pure water.

Will Alternative-Fuel Vehicles Solve the World Energy Crisis?

❝Using innovative, clean technologies available today, we can move beyond our reliance on dirty and unsafe energy sources and our dependence on unstable regions of the world.❞

—Natural Resources Defense Council, "A Responsible Energy Plan for America," 2005. www.nrdc.org.

The Natural Resources Defense Council is an environmental organization.

❝There is currently no . . . advantage to consumers in buying a FFV [flexible-fuel vehicle].❞

—*Consumer Reports*, "The Ethanol Myth," October 2006, p. 16.

Consumer Reports magazine is published by the Consumers Union.

* Editor's Note: While the definition of a primary source can be narrowly or broadly defined, for the purposes of Compact Research, a primary source consists of: 1) results of original research presented by an organization or researcher; 2) eyewitness accounts of events, personal experience, or work experience; 3) first-person editorials offering pundits' opinions; 4) government officials presenting political plans and/or policies; 5) representatives of organizations presenting testimony or policy.

"No country has done more to pioneer . . . ethanol than Brazil."

—Thomas L. Friedman, "The Energy Harvest," *New York Times*, September 15, 2006, p. A25.

Friedman is a *New York Times* columnist.

"Fuel ethanol should be viewed as a sound energy alternative. Its main characteristics are in line with sustainable development needs."

—Alfred Szwarc, "Ethanol Usage in Automotive Fuels," Clean Air Initiative in Latin American Cities, August 15, 2001. www.cleanairnet.org.

Szwarc is an ethanol consultant in São Paulo, Brazil.

"Energy independence is our shared national goal. . . . [We] can meet that goal with biofuels from America's farms—and help the environment, the economy, and our nation's cherished agricultural tradition in the process."

—Tom Daschle, "Follow the Farmers," *American Prospect*, April 2006, p. A17.

Daschle is a former senator from South Dakota.

"The point of flexible-fuel vehicles . . . [is] to reduce dependence on oil and to redirect the money . . . to American agriculture interests and away from often-hostile foreign fuel suppliers."

—James R. Healey, "You Might Be Driving a Flexible Fuel Vehicle," *USA Today*, May 4, 2006. www.usatoday.com.

Healey is a *USA Today* reporter.

66Ethanol production in the United States does not benefit the nation's energy security, its agriculture, the economy, or the environment.99

—David Pimentel and Tad Patzek, "Study Says Ethanol Isn't Worth the Energy," *Journal of Soil and Water Conservation*, November/December 2005, p. 142.

Pimentel is an ecology professor at Cornell University and Patzek is an engineering professor at the University of California–Berkeley.

66Biodiesel is by far our best alternative fuel option at present. . . . It is renewable, sustainable, & domestically produced.99

—Jules Dervaes, "Projects: Backyard Biodiesel," Path to Freedom, March 9, 2005.www.pathtofreedom.com.

Dervaes is an environmentalist and supporter of biodiesel fuel technology.

66While the fossil-fuel era enters its sunset years, a new energy regime is being born that has the potential to remake civilization along radically new lines—hydrogen.99

—Jeremy Rifkin, "Hydrogen: Empowering the People," *Nation*, December 23, 2002, p. 20.

Rifkin is the author of 16 books on the economy, society, and the environment.

"It's unlikely people will wish to drive with a high temperature [hydrogen fuel cell] under their seats, or locate large stores of [explosive] fuel down at the station at the corner."

—Donald R. Sadoway, "Fuel Cells and Portable Power Solutions," September 25, 2006. http://mitworld.mit.edu.

Sadoway is a professor of materials chemistry at the Massachusetts Institute of Technology.

"Significant breakthroughs will be required to lower the cost of hydrogen for it to be competitive against the ever-improving performance of the most advanced internal combustion engine and hybrid technologies."

—Lee R. Raymond, "Challenge, Opportunity & Change: The New Frontiers of Energy," ExxonMobil, June 4, 2003. http://exxonmobil.com.

Raymond is the former chairman and CEO of ExxonMobil.

Will Alternative-Fuel Vehicles Solve the World Energy Crisis?

- It costs about $700 to convert a standard diesel engine to run on 100 percent vegetable oil.

- In 2006 more than 212,000 hybrid cars, trucks, and SUVs were traveling on American roads.

- In 2005, 62 percent of all new cars sold in the world were flex-fuel cars that could run on either ethanol, gasoline, or any combination of the two.

- In Brazil 34,000 gas stations offer ethanol compared with around 700 in the United States.

- By 2050 biofuel production in the United States could reach the equivalent of 7.9 million barrels of oil per day.

- Over 140 billion gallons of biodiesel could be produced in algae ponds covering about 12.5 percent of the Sonoran Desert.

- In 2005 the first hydrogen fuel cell for a vehicle was made available to consumers.

- Hybrids like the Toyota Prius, Honda Civic Hybrid, Ford Escape Hybrid, and the Toyota Camry hybrid have 40 to 80 percent better fuel economy than similar nonhybrid models.

Will Alternative Fuel Vehicles Solve the World Energy Crisis?

How Hydrogen Can Be Used in Vehicles

The only emission from a hydrogen fuel cell tank is water, which is released as steam. Some argue that because hydrogen is highly flammable, using these cells in automobiles is dangerous. The cells are also extremely expensive, costing hundreds of thousands of dollars each.

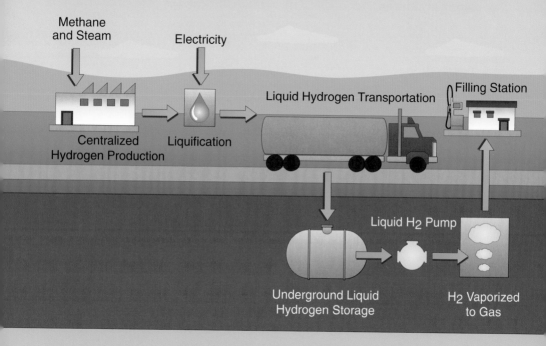

Methane and Steam

Electricity

Liquid Hydrogen Transportation

Filling Station

Centralized Hydrogen Production

Liquification

Liquid H_2 Pump

Underground Liquid Hydrogen Storage

H_2 Vaporized to Gas

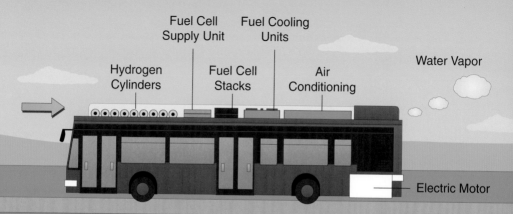

Fuel Cell Supply Unit

Fuel Cooling Units

Hydrogen Cylinders

Fuel Cell Stacks

Air Conditioning

Water Vapor

Electric Motor

Source: Transport for London, "Fuel Cell Buses on Route 25," www.ttl.gov.uk.

Alternative-Fuel and Hybrid Vehicles in Use

With increasing gas prices, concerns about rising pollution, and new environmental regulations, more people are considering alternative-fuel or hybrid vehicles. Auto companies have responded to these factors by offering consumers some options. In 2005 there were millions of flexible-fuel vehicles in use.

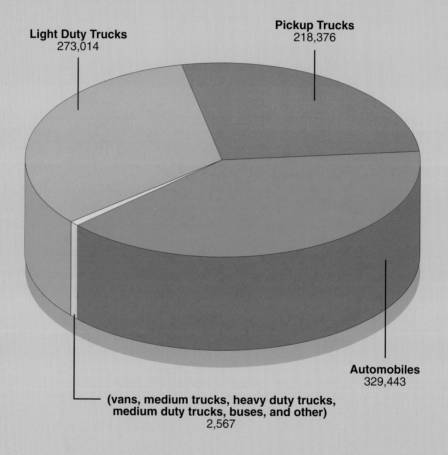

Light Duty Trucks
273,014

Pickup Trucks
218,376

Automobiles
329,443

(vans, medium trucks, heavy duty trucks, medium duty trucks, buses, and other)
2,567

Source: Energy Information Administration, "Alternative Fueled Vehicles Made Available," September 2005. www.eia.doe.gov.

- Gas-electric hybrids constitute barely 1 percent of the nearly 17 million new cars and trucks sold annually in the United States.

- The 2007 BMW Hydrogen 7 Saloon emits only water vapor when running on hydrogen.

Alternative-Fuel Vehicles in the United States, by Fuel, 1998–2004

This graph shows the use of alternative fuels in the United States between 1998 and 2004. Liquid petroleum gas is the most widely used alternative fuel for vehicles, but the number of electric cars in use is increasing at a much faster rate.

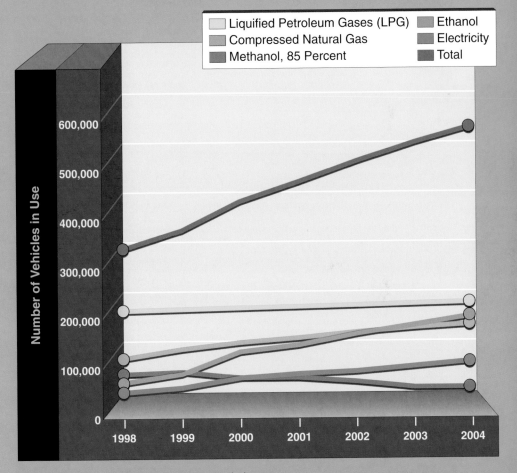

Source: Energy Information Administration, 2004. www.eia.doe.gov.

49

Benefits and Challenges of Alternative Fuels

This table shows the advantages and disadvantages of using alternative fuels. Many people are not aware of the many challenges alternative fuels present.

FUEL	BENEFITS	CHALLENGES
Biodiesel	- domestically produced from nonpetroleum, renewable resources - can be used in most diesel engines, especially newer ones - less air pollutants (other than nitrogen oxides) and greenhouse gases - biodegradable - nontoxic - safer to handle	- use of blends above B5 not yet warrantied by auto makers - lower fuel economy and power (10% lower for B100, 2% for B20) - currently more expensive - more nitrogen oxide emissions - B100 generally not suitable for use in low temperatures - concerns about B100's impact on engine durability
Ethanol	- domestically produced, reducing use of imported petroleum - lower emissions of air pollutants - more resistant to engine knock - added vehicle cost is very small	- can only be used in flex-fuel vehicles - lower energy content, resulting in fewer miles per gallon - limited availability - currently expensive to produce
Hydrogen	- domestically produced, reducing use of imported petroleum - no air pollutants or greenhouse gases when used in fuel cells; produces only NO_x when burned in ICEs	- currently expensive to produce and find (most locations in CA) - currently very expensive - limited availability - can only store enough hydrogen to travel 200 miles
Natural Gas	- 87% is produced domestically - 60%–90% less smog-producing pollutants - 30%–40% less greenhouse gas emissions - less expensive than gasoline	- limited availability - less readily available than gasoline and diesel - fewer miles per gallon
Propane	- fewer toxic and smog-producing pollutants - 85% of LPG used in the U.S. comes from domestic sources - less expensive than gasoline	- no new passenger cars or trucks commercially available (vehicles can be retrofitted for LPG) - less readily available than gasoline and diesel - fewer miles per gallon

Source: United States Department of Energy, "Alternative Fuels," 2006. www.fueleconomy.gov.

Can Nuclear Power Supply the World's Energy Needs?

> 66 When we reach the end of the age of fossil fuels the world will have to consider increasing the use of nuclear power. 99
>
> —David Goodstein, *Out of Gas*, 2004.

Every day, nuclear power plants generate electricity in 31 nations as diverse as Finland, Bulgaria, China, France, and the United States. All of these countries share one thing in common—they use nuclear power plants to lower their reliance on fossil fuels. For example, Japan, which has to import nearly 100 percent of its oil and natural gas, has constructed 55 nuclear reactors since 1970. In 2006 these plants produced about 30 percent of the nation's electricity. A similar situation exists in France where a lack of fossil fuels has encouraged planners to build 58 power plants since 1973. In 2006 nuclear power accounted for 78 percent of the electricity in France, a higher percentage than in any other nation.

Sixteen other countries, including the United Kingdom, the United States, Spain, and Russia rely on nuclear power plants for at least 20 percent of their electricity; however, in these nations, about 80 percent of the nuclear power plants are over 15 years old and there are few plans to build new ones for several reasons. Nuclear power plants cost about 60 percent more than those fueled with coal or natural gas. Because fossil fuels have been readily available in Europe and the United States, most utility companies have not been interested in investing in nuclear power.

Nuclear Accidents

Besides expense, electric utilities face widespread public resistance concerning construction of nuclear power plants. Much of this opposition has to do with two major nuclear accidents. On March 29, 1979, there was a partial meltdown in one of the reactors at the Three Mile Island (TMI) power plant near Harrisburg, Pennsylvania. Hundreds of thousands of people were forced to evacuate the area, and fear of nuclear disaster swept across the nation as the problem went unsolved for several days. In the aftermath, public opinion solidified against nuclear power, and no commercial nuclear reactor has been built in the United States since that time.

> "Because fossil fuels have been readily available in Europe and the United States, most utility companies have not been interested in investing in nuclear power."

More than 25 years later, Americans continue to reject nuclear power as a way of lessening reliance on fossil fuels. A 2006 *Los Angeles Times*/Bloomberg poll asked adults nationwide the best way for the United States to reduce reliance on oil. Only 6 percent supported nuclear power while 52 percent supported developing alternative energy sources.

Nuclear Phase Outs

Another nuclear accident, at the Chernobyl Power Plant in Ukraine, helped galvanize European opinion against nuclear power. In 1986 a major explosion at the plant released a toxic plume of nuclear fallout that drifted across Europe and then around the globe. Fifty-six people died immediately. Approximately 9,000 later became very ill or died. Nearly half a million people in the surrounding vicinity were evacuated and much of the region remains unsafe for human habitation today.

In the aftermath of Chernobyl many European nations, including Belgium, Germany, Sweden, Italy, and the Netherlands held referendums on nuclear power. Voters agreed to slowly phase out nuclear power, shutting down existing plants as they aged. By 2006, however, only one German nuclear power plant had been shut down.

If this plant remained open, however, Germany might not have faced a crisis in January 2006 when natural gas supplies from Russia were interrupted because of a political dispute. Since Germany receives 30 percent of its natural gas from Russia, there were fears of blackouts before deliveries resumed. Commenting on the situation, politician Markus Söder told a German television station, "If we rely too much on oil or gas—for example, Russian gas—then we can run into massive problems. In the long term there is no alternative to nuclear power."[16]

A Growing Demand in Asia

Outside of Europe, nations such as China and India face fossil-fuel shortages and rapidly increasing energy demands. Consequently, the public has been willing to overlook the negative aspects many Westerners associate with nuclear power. This is especially true in China, which had the world's fastest-growing economy in 2006 and where power shortages commonly result in blackouts.

China currently relies on coal to generate 75 percent of its power; however, most of the nation's coal reserves are located in the north part of the country. Because of its aging railroads, China cannot efficiently deliver the coal to the southern coastal areas where there is the most economic growth. To solve this problem, China is building or planning to build 30 nuclear reactors, mostly in the south, before 2026. While China leads the world in nuclear construction, studies have shown that the nation will need at least 300 more nuclear plants by 2050 to meet growing demands.

India is another nation with growth patterns similar to China. Not only is the Indian economy expanding rapidly but the nation's population of 1 billion is expected to increase to 1.5 billion by 2020. This will triple India's energy demands in the next two decades. To deal with the increase, Indian planners are in the process of building nine nuclear plants with many more slated for construction. By the middle of the 21st century, India plans to increase its nuclear generating capacity 10,000 percent.

> By the middle of the 21st century, India plans to increase its nuclear generating capacity 10,000 percent.

Environmental Benefits

Supporters of nuclear power see more plant construction in India and China as a positive trend since both nations will be reducing their reliance on coal, a major source of pollution. And nuclear power plants have a lower environmental impact for other reasons. Some of the benefits have to do with land and habitat preservation, according to the Nuclear Energy Institution (NEI), a trade group:

> Because nuclear power plants produce a large amount of electricity in a relatively small space, they require significantly less land for siting and operation than all other energy sources. For instance, solar and wind farms must occupy substantially more land, and must be sited in geographically unpopulated areas far from energy demand.[17]

For a solar farm to generate as much electricity as a nuclear reactor would require about seven times more land. The equivalent wind farm would require 300 times more space than a single nuclear plant.

What About Nuclear Waste?

Despite industry claims about clean, safe energy, the long-term problems of radioactive waste disposal have yet to be solved. With 104 nuclear reactors, more than any other nation, the United States provides a good example of the waste dilemma. As of January 2004 there were 54,000 tons of uranium waste in the United States produced by commercial nuclear reactors. Ninety-eight percent of the waste is stored at 68 sites around the country, much of it at the power plants where it was generated.

Since a portion of the radioactive waste remains extremely poisonous for at least 30,000 years, Congress has authorized billions of dollars since 1987 to entomb it in the Yucca Mountain Repository in Ney County, Nevada. The site was chosen because the mountain is formed from extremely dense, compacted volcanic ash called tuff. Supporters of the repository say that this makes the mountain extremely stable—it will not erode, collapse, or otherwise degrade for thousands of years while the nuclear waste is buried within it. Opponents of the plan, however, have shown that Yucca Mountain is full of cracks and fissures. These fractures might provide a path for extremely hazardous radioactive waste to seep into the water table that currently supplies Las Vegas and other cities. In addition, there have been over 600 small and medium earthquakes within 50 miles of the repository since 1976. A major earthquake could be disastrous if it opened large crevasses beneath the repository while cracking apart the concrete casks full of nuclear waste.

Because of these problems dozens of lawsuits have been filed by environmental groups and the state of Nevada in order to stop the U.S. Department

> **Ninety-eight percent of [nuclear] waste is stored at 68 sites around the country, much of it at the power plants where it was generated.**

of Energy from opening the site. While fighting the lawsuits, the government continues with plans to begin receiving waste at Yucca Mountain in 2017—30 years after the site was first chosen. Skeptics believe that the date is overly optimistic and it might take until 2030 to open the repository.

A "Final Resting Place" in Finland

As the United States struggles with Yucca Mountain, scientists in Finland are pushing ahead with a plan to bury nuclear waste at the Onkalo facility by 2020. Engineers at Onkalo have drilled a tunnel 20 feet wide and 1,600 feet deep through solid rock. They plan on encasing spent fuel rods from nuclear reactors in corrosion-resistant copper canisters. These would be deposited in thousands of shafts drilled at the bottom of the tunnel.

By 2100 the repository will be filled and the tunnel mouth will be sealed leaving no sign of the deadly waste contained below. With plans at Onkalo progressing, BBC correspondent Richard Black writes, "Finland is on course to become the first country in the world to entomb its most troublesome nuclear waste in a designated final resting place."[18] Environmentalists disagree with this notion, however, because they do not believe in the safety of burying nuclear waste. As Greenpeace spokesperson Kaisa Kosonen states: "I would like to see much more research done and . . . I would not want this marketed as 'waste issue solved,' because it's not."[19]

A New Generation of Nuclear Plants

Because of issues concerning hazardous waste and nuclear safety, scientists have been working to develop a new generation of nuclear power plants. These plants are designed to be more efficient, generate less waste, and have advanced controls to prevent accidents. The goal is to lower costs and alleviate public concerns so that nuclear power can someday replace the electricity currently generated by fossil fuels.

The new nuclear designs are called Generation IV because they have evolved from the Generation III plants built in the 1970s and 1980s that continue to operate today. The Generation IV reactors will employ six different types of complex technologies. Some of them will produce hydrogen as a by-product of nuclear fission. This gas would be used to supply fuel for hydrogen-powered vehicles.

Generation IV power plants include passive safety systems overseen by computers. These are designed to eliminate the human error that caused the accidents at Three Mile Island and Chernobyl. In addition, the new plants will be smaller and more economical to build, operate, and maintain than are current nuclear reactors.

Meanwhile, the Chinese are making plans to market a relatively small nuclear generator, called a pebble-bed reactor. The reactor will be about one-fifth the size of a traditional nuclear power plant, be made from mass-produced parts, and be cheap enough for private customers to purchase. It is supposedly accident proof and produces hydrogen as a by-product of fusion. Chinese engineers believe that within a decade small cities or even large factories will be able to meet their energy needs with pebble-bed reactors.

> " The Chinese are making plans to market a relatively small nuclear generator, called a pebble-bed reactor. "

An Ongoing Debate

Even as technological advances make nuclear power more attractive, it remains an energy source with vociferous critics. According to the environmental group Greenpeace, "[Building] enough nuclear power stations to make a meaningful reduction in [fossil-fuel consumption] would cost trillions of dollars, create tens of thousands of tons of lethal high-level radioactive waste, contribute to further proliferation of nuclear weapons materials, and result in a Chernobyl-scale accident once every decade."[20]

Supporters of nuclear power dismiss such claims, saying that nuclear power is the only way to reduce reliance on fossil fuels. And as long as small amounts of uranium generate massive quantities of electricity, there is little doubt that nuclear power will continue to be a major part of the world's energy picture for some time to come.

Primary Source Quotes*

Can Nuclear Power Supply the World's Energy Needs?

66Nuclear power plants have been and are even more so now among the most well-protected elements of our national civilian infrastructure.99

—Nils J. Diaz, "NRC Chairman Discusses Nuclear Plant Security at Americas Nuclear Energy Symposium," Nuclear Regulatory Commission, October 4, 2004. www.nrc.gov.

Diaz is chairman of the U.S. Nuclear Regulatory Commission.

66Each nuclear plant, through accident or malice, could release enough radioactivity to hazard a continent.99

—Amory B. Lovins and L. Hunter Lovins, "The Nuclear Option Revisited," *Los Angeles Times*, July 8, 2001.

The Lovins are longtime advisers to the energy industry.

* Editor's Note: While the definition of a primary source can be narrowly or broadly defined, for the purposes of Compact Research, a primary source consists of: 1) results of original research presented by an organization or researcher; 2) eyewitness accounts of events, personal experience, or work experience; 3) first-person editorials offering pundits' opinions; 4) government officials presenting political plans and/or policies; 5) representatives of organizations presenting testimony or policy.

66The 1986 Chernobyl disaster spewed more radiation across Europe than was released [by the 1945 atomic bomb blast] in Hiroshima . . . and left a region that had been inhabited by 100,000 people a glow-in-the-dark no-man's land.99

—Joshua Holland, "Bush's Nuclear Madness," alternet.com, May 2, 2006. www.alternet.org.

Holland is an Alternet staff writer.

66The United States must increase its capacity to generate energy. . . . Nuclear power should be a significant part of the solution.99

—Eric P. Loewen, "Nuclear Power Can Help Solve Energy Crisis," *National Defense*, August 2001. www.nationaldefensemagazine.org.

Loewen is a nuclear engineer with the U.S. Department of Energy.

66Through the release of atomic energy, our generation has brought into the world the most revolutionary force since the prehistoric discovery of fire.99

—Albert Einstein, "Emergency Committee of Atomic Scientists," Federation of American Scientists, 2006. www.fas.org.

Physicist Einstein's theories helped scientists develop nuclear energy.

66Nuclear power is the only energy source that can be developed on a massive scale that will meet all the requirements for tremendous increases in generation.99

—Dennis Beller, "Atomic Time Machines: Back to the Nuclear Future," American Energy Independence, 2005. www.americanenergyindependence.com.

Beller is a nuclear power plant designer and engineer.

66Today, only two large-scale methods of generating electricity are available to us: burn the fossil fuels— coal, oil, and natural gas—or use nuclear power.99

—Robert C. Morris, *The Environmental Case for Nuclear Power.* St. Paul: Paragon House, 2000, p. ix.

Morris is a chemistry teacher and author.

66To change how we power our homes and offices, we will invest more in . . . clean, safe nuclear energy.99

—George W. Bush, "State of the Union Address by the President," White House, January 31, 2006. www.whitehouse.gov.

Bush is the 43rd president of the United States.

66Yucca Mountain scientists will readily tell you that the question is not if the repository will release its [radio-active] contents, but when.99

—"Don't Dump Nuclear Waste at Yucca Mountain," Greenaction, 2005. www.greenaction.org.

Greenaction is an environmental group.

66Any strategy for India to achieve energy security must involve a mix of many different energy sources, from clean coal, oil and gas to renewables such as wind and solar power—*and* nuclear energy.99

—Ashok Parthasarath, "India's Energy Mix Needs a Nuclear Boost," SciDevNet, May 11, 2006. www.scidev.net.

Parthasarath is the former science adviser to the late Indian prime minister Indira Gandhi.

❝Generation IV nuclear power plants are ... designed to be highly economical, and will minimize waste while providing enhanced safety.❞

—Margaret W. Hunt, "Nuclear Power," *Advanced Materials & Processes,* June 2006, p. 3.

Hunt is editor of *Advanced Materials & Processes* magazine.

....

❝The amount of waste produced in nuclear-generated electricity is vastly less than in fossil fuels.❞

—Scott W. Heaberlin, *A Case for Nuclear-Generated Electricity: (Or, Why I Think Nuclear Power Is Cool and Why It Is Important That You Think So Too).* Columbus, OH: Battelle, 2004, p. 3.

Heaberlin is a nuclear engineer and author.

....

Facts and Illustrations

Can Nuclear Power Supply the World's Energy Needs?

- In 2006 about 17 percent of the electricity generated worldwide was produced by nuclear power plants.

- In 2004 there were 104 nuclear power plants in the United States.

- In 2006 France produced 78 percent of its electricity with nuclear reactors.

- Because of public opposition to nuclear energy, Germany plans to shut down its 20 nuclear power plants by 2026.

- Energy demands in India are expected to triple in the next two decades.

- Nuclear power plants are about 60 percent more costly to build than coal or natural gas plants.

- As of January 2004, there was 54,000 tons of nuclear waste in the United States produced by commercial nuclear reactors.

- The Yucca Mountain Repository is scheduled to begin accepting nuclear waste on March 31, 2017.

The United States Generates the Most Nuclear Power

As of December 2005 there were 443 operating nuclear power plants in the world. France relies on 58 reactors to generate about 78 percent of its electricity, while the United States relies on nuclear-generated power for 20 percent of its electricity.

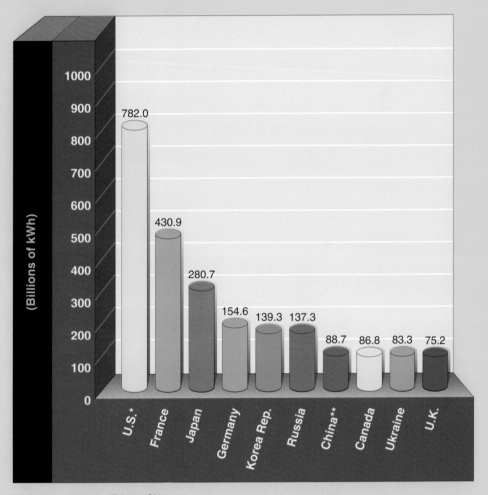

*Preliminary, **Includes Taiwan, China

Source: International Atomic Energy Agency and Global Energy Decisions/Energy Information Administration, April 2006.

63

Operating Nuclear Power Plants in the United States

This map shows the number of operating nuclear power plants in each state. Illinois has the most power plants with six.

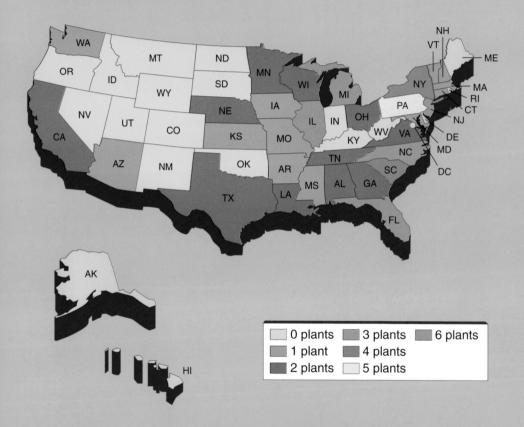

0 plants	3 plants	6 plants
1 plant	4 plants	
2 plants	5 plants	

Source: Energy Information Administration, "US Nuclear Power Plants by State." www.eia.doe.gov.

- Scientists are developing at least six new types of nuclear reactors to be deployed between 2010 and 2030.

- The pebble-bed reactor designed in China will be about one-fifth the size of a traditional nuclear power plant and be constructed from mass-produced parts.

World Nuclear Reactor Construction Is Decreasing

Hundreds of nuclear power plants were constructed around the world between 1950 and the late 1970s. In 1979 a core reactor at the Three Mile Island power plant near Harrisburg, Pennsylvania, melted down. The accident signaled the end of nuclear power plant construction in the United States. Of the 129 plants scheduled for construction, 127 were canceled.

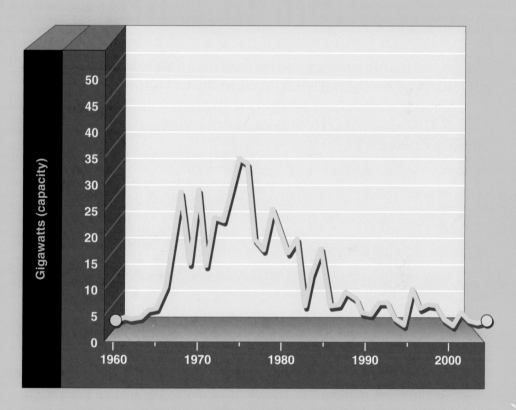

Source: Christopher Flavin, "Nuclear Revival? Don't Bet on It!" *World Watch*, July/August 2006, p.19.

Can Renewable Energy Meet Future Demands for Electricity?

> **" Renewable energy ... is beginning to attract the same kind of buzz that surrounded John D. Rockefeller's feverish expansion of the oil industry in the 1880s—or Bill Gates's early moves in the software business in the 1980s."**
>
> —Christopher Flavin, president of the Worldwatch Institute.

In recent years growing fears over oil prices, energy shortages, and global warming have intensified interest in renewable energy. These so-called green technologies consist of a wide range of power-generating methods that use the sun, wind, biomass, and other sources. All have one thing in common, according to the United Kingdom Renewable Energy Advisory Group: They are relatively nonpolluting "energy flows that occur naturally and repeatedly in the environment and can be harnessed for human benefit."[21]

Proponents of renewables believe that green technology is the only way to prevent an international energy crisis in the future. As German development minister Heidemarie Wieczorek-Zeul told an audience at an alternative energy conference in 2004, "We must be clear that, given the huge demand [for electricity] in developing countries, the world will need far more energy than today. So an enormous expansion of renewable energies is needed."[22]

Technologies that harvest energy from the sun, wind, and earth currently generate about 6 percent of electricity produced worldwide. And according to the UN-affiliated World Renewable Energy Congress (WREC), that number is expected to increase to 70 percent by 2070.

If the WREC forecast is accurate, the world will drastically reduce its reliance on fossil fuels and nuclear power by the end of the century; however, some experts believe that switching to alternative technologies on a massive scale will, ironically, require a significant portion of the world's remaining oil. This point is explained by environmental author Jim Bell: "When [fossil fuel supplies begin] to shrink it is harder to harness the resources necessary to manufacture the solar panels, the wind mills, and the other equipment needed when we begin the inevitable task of creating a large scale alternative infrastructure."[23]

The Future of Biomass

Whatever its role in the future, the renewable energy industry was drawing considerable attention in 2006 as oil and natural gas prices surged. On Wall Street, financiers were investing billions in companies that manufactured wind turbines, solar cells, and biomass technology. Even the chemical manufacturing giant DuPont was getting into the biomass business by searching for ways to convert entire corn plants and other plant material into energy. Commenting on the program, company general manager John Ranieri said, "The world needs new sources of clean energy now."[24]

> Switching to alternative technologies on a massive scale will, ironically, require a significant portion of the world's remaining oil.

Companies like DuPont have focused on biomass because it is big business. Ethanol and biodiesel production accounted for about 50 percent of all renewable energy production in 2006. But there are other, less well-known forms of bioenergy being used today. These are waste products of modern society. They include landfill gas, construction waste, wastepaper, sawdust, pulp from paper mills, and fermented farm waste.

Producing biomass from industrial waste products helps to recycle garbage that would otherwise take up space in landfills. Using this waste also benefits the environment because it is produced and used locally. For example, many companies that manufacture paper and wood products have invested in onsite biomass burners that allow them to convert their wood waste into steam and electricity. This means the manufacturer does not need to pay for waste hauling and does not need to purchase electricity produced with fossil fuels.

A second method of utilizing biomass concerns gasification, in which solid waste is exposed to heat and oxygen to produce methane gas. Sometimes called landfill gas or biogas, this process recycles municipal solid waste (MSW); that is, trash that comes from food scraps, lawn clippings, leaves, and animal waste. To capture the gas, a pile of garbage is injected with water, which accelerates decomposition. As the rotting garbage releases methane, it is captured by an extensive network of pipes and sent to a nearby burner where it is used to produce electricity. One of the most technologically advanced biomass generators is found in Woodlawn, Australia, southwest of Sydney. This bioreactor receives about 441,000 tons of garbage annually and produces enough power for about 13,000 homes.

Plants like the one in Woodlawn can help reduce fossil-fuel consumption, but such bioreactors produce less than 1 percent of all electricity worldwide. However, experts say that if all biomass resources were utilized for energy, that number could increase tenfold.

Energy from Wind

Unlike the complicated systems associated with biomass, wind turbines are relatively simple and inexpensive. These machines generate electricity when the wind blows. The movement of their two- or three-blade rotors is transmitted to a generator that produces electricity. This flows through heavy cables into a transformer, into power lines, and on to consumers.

A single wind turbine can power about 80 average homes, but most electric utilities utilize hundreds or even thousands of machines on a centralized "wind farm" located in an extremely windy area. The largest wind farms are in California at Altamont Pass, San Gorgonio Pass, and Tehachapi Pass, where a total of 13,000 wind turbines supply power for Central and Southern California. Together, California's wind farms produce about 1.5 percent of the state's total electricity, more than enough to light a city the size of San Francisco.

Elsewhere in the world, wind farms produce a greater percentage of power. In Germany and Spain consumers receive over 6 percent of their power from wind, while in Denmark 20 percent of the nation's electricity is generated by wind turbines. Other nations that are increasingly relying on wind power include India, the Netherlands, and the United Kingdom.

Criticism of Wind Power

According to a study by the U.S. State Department, wind power is the fastest-growing new source of electricity worldwide because it is the cleanest and cheapest renewable energy source. But there has been public opposition to large-scale wind projects in many places throughout the world. One wind energy critic, Eric Rosenbloom, describes problems associated with the turbines:

> Because of the intermittency and variability of the wind, conventional power plants must be kept running at full capacity to meet the actual demand for electricity. They cannot simply be turned on and off as the wind dies and rises, and such inefficient operation would actually increase their output of pollution and carbon dioxide.[25]

In addition, it would take millions of wind towers to fill the growing demand for electricity across the globe. For example, if the United States were to increase its proportion of wind-generated electricity from 1 to 5 percent it would require the construction of almost 2 million wind towers. Such an ambitious objective would doubtless face public resistance. People who live close to wind towers complain that the rotating blades produce an intolerable level of noise. According to Rosenbloom, there is a "penetrating low-frequency aspect to the noise, a thudding vibration, much like the throbbing bass of a neighboring disco, that travels much farther than the usually measured 'audible' noise . . . and people have complained that it causes anxiety and nausea."[26]

> "California's wind farms produce about 1.5 percent of the state's total electricity, more than enough to light a city the size of San Francisco."

Unlimited Power from the Sun

While wind farms are problematic because of their large scale, solar-powered photovoltaic (PV) cells are small enough to install on the roof of a single home. For example, a small, two-kilowatt solar energy system can supply an average-size home with up to 80 percent of its electrical needs, depending on the number of lights, appliances, air conditioners, heaters, and electronics that are used.

Because PV cells are easy to buy, install, and use, the photovoltaic industry has been growing at an accelerated pace. Between 2001 and 2005 worldwide production of solar cells increased by 60 percent. In the United States alone, the PV industry is a $2 billion business that employs about 20,000 people.

Solar Opponents

As with other renewable energy sources, solar power has problems that limit its use. The biggest obstacle is that the sun is only available for use during the day while the greatest demand for electric power is for illumination at night. In addition, cloudy skies, snow, and dust dim the ability of solar collectors to produce energy efficiently. With no way to effectively store the massive amount of energy required to light cities during storms or after dark, some believe solar power is an impractical alternative to fossil fuels.

Large-scale, centralized solar plants also have extremely high maintenance costs. For example, a solar facility built in the sunny Mojave Desert near Barstow, California, cost approximately $200 million. The giant solar reflectors at the site, about the size of 15 football fields, needed to be cleaned once a week. Other maintenance and repair issues added to the plant's overhead. Meanwhile, the facility was able to generate only about $1.7 million dollars' worth of energy a year, less than 1 percent of construction costs. In the end, the government agencies and corporate partners that operated the project lost so much money they had to shut the solar plant down.

> **Large-scale, centralized solar plants also have extremely high maintenance costs.**

There are other problems associated with large-scale solar plant construction, according to physicist and engineer Mike Oliver and emeritus professor John Hospers:

> Putting the entire U.S. on a solar electric regimen would require an initial expenditure of at least $30 trillion, plus trillions more each year for upkeep and maintenance. Enthusiasts also overlook the fact that we would need more than 500 times as much construction material (copper, iron, concrete, steel, etc.) to build solar generators as would be required for conventional power plants. . . . Entire mountain ranges would have to be ripped up to provide the staggering amount of gravel, asphalt, cement, metals, etc. required.[27]

New Solar Technology

In 2005 there was a new development in solar technology that researchers hope will overcome the problems mentioned by Oliver and Hospers. Inventors developed a solar collection system that uses an advanced manufacturing technique called nanotechnology to create solar cells that are 100,000 times smaller than the width of a human hair. The nanocells are printed onto thin rolls of flexible plastic film. Sheets of the film can collect more light than traditional PV cells of the same size. Manufacturing costs are about one-tenth of traditional silicon cells and the sheets of film can be spread across any rooftop or built into roofing materials.

Those familiar with the technology say that the film will be available for consumer use by 2010. The first application of the film will likely be in cell phones and other electronic devices, replacing rechargeable batteries like those used today. Some believe, however, that the nano–solar film could replace all fossil-fuel power plants in the United States within 30 years. According to Paul Carlstrom in the *San Francisco Chronicle,*

> both inventors and investors are betting that flexible sheets of tiny solar cells used to harness the sun's strength will ultimately provide a cheaper, more efficient source of energy than the current smorgasbord of alternative and fossil fuels. Solar energy could furnish much of the

nation's electricity if available residential and commercial rooftops were fully utilized.[28]

Making the Switch

Nano–solar cells are just one of the futuristic alternative energy sources being developed today. Others have hopes of harnessing the energy in ocean waves or even placing giant solar collectors in outer space. Those who support such projects believe that society must switch to alternative sources before a world energy crisis causes widespread chaos.

> **Renewable sources available today can provide only a fraction of the power that the modern world needs.**

While the renewable sources available today can provide only a fraction of the power that the modern world needs, improvements in technology will likely allow civilization to one day end its reliance on oil, coal, and natural gas. Whatever the ultimate odds of success, millions are already switching to solar panels, windmills, and other alternative energy sources. And those systems will provide electricity to power-hungry consumers as long as the wind blows and the sun shines.

Primary Source Quotes*

Can Renewable Energy Meet Future Demands for Electricity?

"PV technologies, which use semiconductor technology to convert sunlight directly into electricity, are good for our energy, our economy, our environment, and our future."

— "Frequently Asked Questions About Photovoltaics," National Center for Photovoltaics, 2006. www.nrel.gov.

The National Center for Photovoltaics is part of the U.S. Department of Energy.

"Current [solar] technology is inefficient, underdeveloped, and in most cases too expensive to be cost effective."

— Greg Gagliardi, "Solar Power," Renewable Energy Sources, 2004. www.personal.psu.edu.

Gagliardi is an energy researcher.

* Editor's Note: While the definition of a primary source can be narrowly or broadly defined, for the purposes of Compact Research, a primary source consists of: 1) results of original research presented by an organization or researcher; 2) eyewitness accounts of events, personal experience, or work experience; 3) first-person editorials offering pundits' opinions; 4) government officials presenting political plans and/or policies; 5) representatives of organizations presenting testimony or policy.

66Sustainable energy development should provide adequate energy services for satisfying basic human needs, improving social welfare, and achieving economic development throughout the world.**99**

—Howard S. Geller, *Energy Revolution: Policies for a Sustainable Future.* Washington, DC: Island, 2003, p. 16.

Geller is an author and the director of the Southwest Energy Efficiency Project.

....

66Fundamental human goals include both the desire for abundant energy on demand and a clean and safe environment.**99**

—Paul Kruger, *Alternative Energy Resources: The Quest for Sustainable Energy.* Hoboken, NJ: John Wiley & Sons, 2006, p. 1.

Kruger is professor emeritus at Stanford University.

....

66There is enormous potential in biomass, to generate renewable energy, to help the environment and to provide another possible market for our farmers.**99**

—William Bach, "UK: Getting the Best Out of Biomass," oneworld.net, http://us.oneworld.net.

Bach is British minister for sustainable farming and food.

....

66To fix [the energy system], we should look to localized, distributed, democratic energy sources, such as solar (photovoltaic) electricity, and free ourselves from the tyranny of fossil fuels.**99**

—Kate Cell, "Energy Democracy," tompaine.com, August 1, 2006. www.tompaine.com.

Cell is the director of development at the Prometheus Institute for Sustainable Development.

....

Can Renewable Energy Meet Future Demands for Electricity?

"In Europe environmental concerns really drive energy policy which then drives the growth of wind power."

—Randall Swisher, "Ireland to Build World's Largest Wind Farm," *National Geographic Today,* January 15, 2002. http://news.nationalgeographic.com.

Swisher is executive director of the American Wind Energy Association.

"The resulting sound of several [wind turbines] together has been described to be as loud as a motorcycle, like aircraft continually passing overhead, a brick wrapped in a towel turning in a tumble drier."

—Eric Rosenbloom, "A Problem with Wind Power," Kirbymountain.com, January 2004. www.kirbymountain.com.

Rosenbloom is a journalist.

"Concerns with rising energy costs, power disruptions, pollution and global warming have combined with reductions in the cost of PV cells ... to make solar power shine brighter than it has for decades."

—Joe Provey, "The Sun Also Rises—Again; After Decades of Setbacks, Solar Energy Returns and Many Believe That It's Here to Stay," *Popular Mechanics,* September 2002, p. 92.

Provey is a journalist.

"[The] mass of receptors, motors, rotating platforms, and reflectors required for a national solar electric system would occupy some 25,000 square miles of land."

—Mike Oliver and John Hospers, "Alternative Fuels?" *American Enterprise,* September 2001, p. 20.

Oliver is a physicist and engineer and Hospers is an emeritus professor.

❝The dream in solar energy is to develop technology so that someday, your house is like a little generating plant, and if you don't use the power you feed it back into the grid.❞

—George W. Bush, "President Discusses Energy During Visit to Nuclear Generating Station in Pennsylvania," May 24, 2006. www.whitehouse.gov.

Bush is the 43rd president of the United States.

❝While not a silver bullet, the more we use [renewable energy sources], the better off we will be in terms of reducing oil imports, reducing pollution and greenhouse gas emissions, and increasing jobs.❞

—Michael Eckhart, "Renewable Energy Poised to Boom," Off-Grid, July 3, 2006. www.off-grid.net.

Eckhart is president of the American Council on Renewable Energy.

❝Photovoltaic and wind systems can conceivably provide all U.S. electricity needs (with major modification to our grid), eliminating our use of coal, oil and natural gas for electricity generation.❞

—John A. Turner, "The Sustainable Hydrogen Economy," *Geotimes*, August 2005. www.geotimes.org.

Turner is a scientist at the National Renewable Energy Laboratory in Golden, Colorado.

Facts and Illustrations

Can Renewable Energy Meet Future Demands for Electricity?

- In 2005, 6 percent of the electricity used on earth came from renewable sources.

- Biomass reactors generate power from waste products, including landfill gas, construction waste, sawdust, pulp from paper mills, and fermented farm waste.

- In 2001 energy firms spent a record $1.7 billion on wind projects, in the United States, increasing wind power capacity by 60 percent—enough to provide electricity to 1 million homes.

- In the United States adequate winds for commercial power production are found at sites in 46 states.

- North Dakota alone could supply over 40 percent of the nation's electricity if enough wind farms were constructed in that state.

- Every hour the sun radiates more energy onto the earth than the entire human population uses in one year.

- A small, two-kilowatt solar energy system can supply an average-size home with up to 80 percent of its electrical needs.

- During operation, photovoltaic solar cells produce no air pollution, little or no noise, and require no fuels.

Renewable Energy Is a Small Portion of Total Energy

About 6 percent of the world's energy is currently produced by renewable resources. The World Renewable Energy Congress projects that by 2070 70 percent of electricity worldwide will be generated by renewable resources.

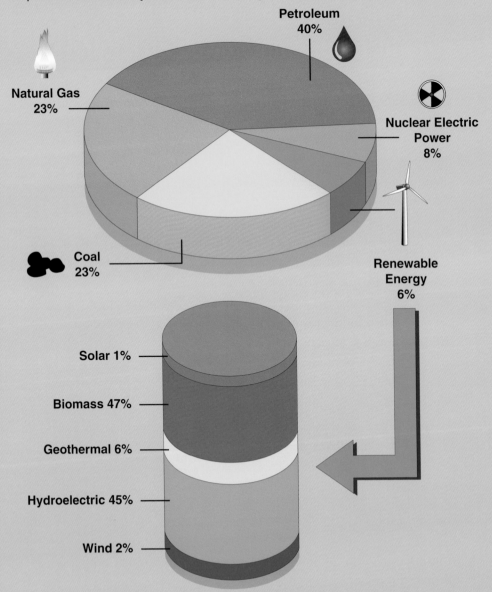

Petroleum
40%

Natural Gas
23%

Nuclear Electric
Power
8%

Coal
23%

Renewable
Energy
6%

Solar 1%

Biomass 47%

Geothermal 6%

Hydroelectric 45%

Wind 2%

Source: Energy Information Administration, "Renewable Energy Trends," September 2005. www.eia.doe.gov.

Some States Require Minimum Levels of Renewable Energy

The World Renewable Energy Congress predicts that by the year 2070 between 60 and 80 percent of all electricity will come from renewable sources. Twenty-two states and the District of Columbia have set legal deadlines for utility companies to generate a certain portion of the states' electricity from renewable resources.

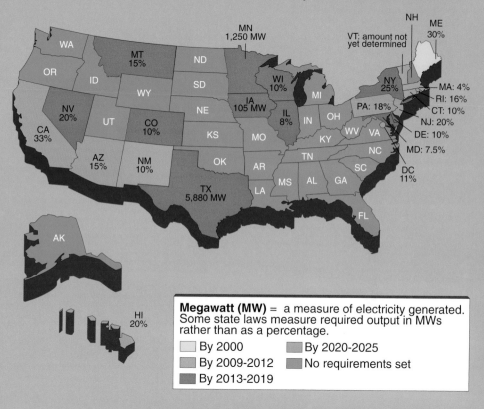

NH
ME 30%
MN 1,250 MW
VT: amount not yet determined
WA
MT 15%
ND
WI 10%
NY 25%
MA: 4%
RI: 16%
OR
ID
SD
MI
CT: 10%
WY
IA 105 MW
PA: 18%
NJ: 20%
NV 20%
NE
IL 8%
IN
OH
DE: 10%
CA 33%
UT
CO 10%
KS
MO
KY
WV
VA
MD: 7.5%
AZ 15%
NM 10%
OK
AR
TN
NC
DC 11%
TX 5,880 MW
MS
AL
GA
SC
LA
FL
AK

Megawatt (MW) = a measure of electricity generated. Some state laws measure required output in MWs rather than as a percentage.

- By 2000
- By 2009-2012
- By 2013-2019
- By 2020-2025
- No requirements set

HI 20%

Source: The Pew Center on Global Climate Change, May 2006. www.pewclimate.org.

- The average home has more than enough roof space for a solar cell that would provide all of its electrical power needs.

- Concentrating photovoltaic systems (CPS) use mirrors or lenses to direct concentrated sunlight onto high-efficiency solar cells.

How Solar Power Is Used on Earth

Each day the sun provides more energy (measured in terawatts [TW]) than would be consumed by the entire population of the earth in 27 years. Unfortunately, solar-powered systems are currently expensive and there is no economical way to store energy for use during cloudy periods.

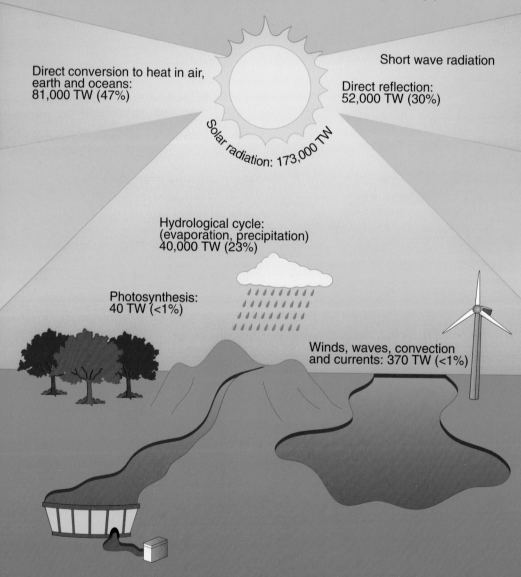

Short wave radiation

Direct conversion to heat in air, earth and oceans: 81,000 TW (47%)

Direct reflection: 52,000 TW (30%)

Solar radiation: 173,000 TW

Hydrological cycle: (evaporation, precipitation) 40,000 TW (23%)

Photosynthesis: 40 TW (<1%)

Winds, waves, convection and currents: 370 TW (<1%)

Source: Godfrey Boyle, *Renewable Energy: Power for a Sustainable Future.* New York: Oxford University Press, 1996.

How Wind Can Create Energy

A wind turbine generates electricity when the wind blows and turns the rotors. The movement of the spinning rotors is transmitted to a generator that produces electricity, which flows through heavy cables into a transformer at the base of the machine. From there the electricity flows into power lines and on to the consumers. Wind power supplies less than 1 percent of energy needs in the United States.

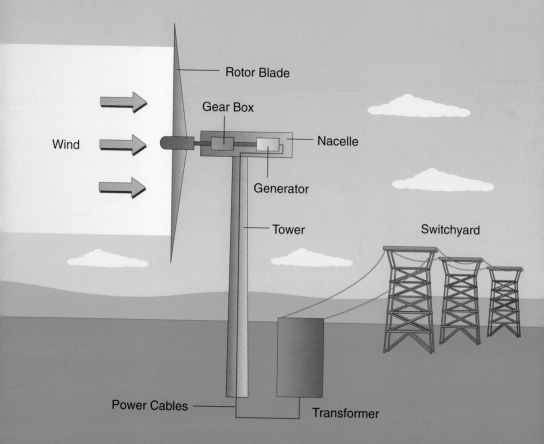

Source: Tennessee Valley Authority, "Wind Turbine Energy: Power Out of Thin Air–It's No Magic Trick," 2006. www.tvs.gov.

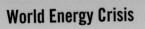

Key People and Advocacy Groups

Sadad al-Husseini: A former head of exploration and production at Aramco, the national oil company of Saudi Arabia, al-Husseini has given many interviews in which he points out that an insufficient number of oil refineries are limiting gasoline production, which will hinder supplies in the future. He also believes world oil production will peak in 2015.

Sir Richard Branson: As a founder of Virgin Group, Branson made his fortune as CEO of Virgin Records, Virgin Air, and the rail network Virgin Trains. In 2006 Branson committed $3 billion to be invested into research and development of new biofuels suitable for both ground and air transportation.

George W. Bush: Before his career in politics, which culminated in his becoming president of the United States in 2000, Bush was an entrepreneur in the Texas oil industry. Between 2000 and 2006 Bush appointed more than 50 former oil company executives to high-level roles in the executive branch, including vice president Richard Cheney, secretary of state Condoleeza Rice, and senior adviser Karl Rove.

Colin J. Campbell: As an exploration geologist for several major oil companies, Campbell has written extensively about peak oil issues. In a widely quoted article in the March 1998 *Scientific American,* Campbell warned that world oil production will peak in 2010.

Julian Darley: A British environmental researcher and author of *High Noon for Natural Gas,* Darley lectures widely on the implications of the coming natural gas peak and the subsequent decline in production.

Kenneth S. Deffeyes: In his 2001 book *Hubbert's Peak: The Impending World Oil Shortage,* Deffeyes refined Hubbert's mathematical theories and explained their meaning in layman's terms. Deffeyes paints a bleak picture of a doomsday where there will be few solutions to the problem of diminishing oil supplies.

Albert Gore: Gore was vice president of the United States between 1992 and 2000. After leaving office he became a dedicated environmentalist, traveling around the world to deliver a keynote presentation about the dire consequences of global warming to students, government organizations, and environmentalists. The presentation was made into the film *An Inconvenient Truth,* which broke box office records for a documentary movie in 2006.

Marion King Hubbert: As a geophysicist who worked for the Shell Oil research lab in Houston, Texas, between 1943 and 1964, Hubbert studied the capacities of oil fields. In 1956 he devised a formula that predicted petroleum production in the United States peaking sometime in the early 1970s and falling off thereafter. His formula, called Hubbert's Peak, proved to be accurate and has been used by those who predict world oil production peaking in the coming years.

Peter Huber: Huber is a critic of the peak oil theory and author of the book *Hard Green: Saving the Environment from the Environmentalists,* in which he argues that humanity will always find ways to capture enough energy for its needs.

Vinod Khosla: As a venture capitalist and founder of Sun Microsystems, Vinod Khosla is now investing his fortune in ethanol fuel research. Khosla wants to move away from corn-based ethanol and develop ethanol fuel based on nonfood plant matter such as grasses and wood chips.

Bjørn Lomborg: As author of *The Skeptical Environmentalist,* Lomborg argues against the peak oil theory by pointing out that there are many as yet undiscovered oil reserves that will provide for future needs. In addition, Lomborg believes that technology will continue to provide ever more efficient methods for recovering oil from known reserves.

David J. O'Reilly: As the chairman and chief executive officer of Chevron, the world's second largest energy company, O'Reilly has repeatedly stated that the era of easy access to petroleum is over. O'Reilly helped institute Chevron's "Will You Join Us" campaign, a series of advertisements that highlight problems concerning oil.

The World Research Institute: The World Research Institute is a Washington, D.C., think tank founded by James Gustave Speth, co-founder of the Natural Resources Defense Council. In 2003 the WRI brought together a dozen major corporations and got them to pledge that by 2010 at least 1,000 megawatts of the power they use will come from renewable sources such as wind, solar, and hydrogen fuel cells. The companies include General Motors Corp., Dow Chemical Co., DuPont, Johnson & Johnson, IBM, Kinko's, and Staples, Inc.

Chronology

1956
Marion King Hubbert, a Shell Oil geologist, formulates a complex mathematical theory known as Hubbert's Peak that predicts United States oil production will peak in the 1970s.

1973
Oil production peaks in Canada. Several Arab nations, angered at U.S. support of Israel in the 1973 Arab-Israeli war, institute an oil embargo against the United States, triggering an oil crisis.

1902
Oldsmobile begins mass producing gasoline-powered cars.

1880
The first oil wells go into mass production.

1908
Henry Ford sets up the Model T Ford assembly line.

1951
On December 20, the first electrical power created by nuclear fission is generated at the Experimental Breeder Reactor in Arco, Idaho.

1970
Oil production peaks in the United States.

| 1800 | 1900 | 1910 | 1920 | 1930 | 1940 | 1950 | 1960 | 1970 |

1892
The diesel engine is invented by German engineer Rudolf Diesel. It is designed to run on a variety of fuels, including peanut oil.

1938
Three chemists working in a laboratory in Berlin split the uranium atom leading to the discovery of nuclear power.

1974
In March, Arab oil ministers, with the exception of Libya, announce the end of the embargo against the United States.

1977
The Powerplant and Industrial Fuel Use Act passes restricting the construction of powerplants that use petroleum or natural gas as their primary fuels. The main purpose of the law is to promote national energy security by encouraging the use of coal and alternative fuels in new electric power plants.

1978
The Iranian Revolution begins resulting in a production loss of 3.9 million barrels of crude oil per day.

1979
On March 29 it is discovered that the core of a nuclear reactor at the Three Mile Island (TMI) power plant near Harrisburg, Pennsylvania, is in the process of a partial meltdown.

1989

March 24, the *Exxon Valdez* oil tanker runs aground in Alaska spilling 240,000 barrels of crude oil into the environment.

1997

The first all-electric cars produced since the 1920s are offered for lease by major auto companies.

2005

Six percent of the primary energy sources on earth come from renewable sources.

2001

The record for the world's deepest oil well is set in December when Chevron drills 34,189 feet in the Gulf of Mexico.

2000

A sharp increase in U.S. energy prices results in rolling blackouts and frequent system emergencies.

1980   1985   1990   1995   2000   2005

1986

The worst disaster in nuclear history takes place in April when there is a meltdown at the Chernobyl Power Plant in Ukraine.

2004

The world's second largest oil complex, the Cantarell field in Mexico, peaks in production. General Motors cancels its electric car program.

2003

In order to prevent market abuse and provide clear "rules of the road" for all market participants, the Federal Energy Regulation Commission proposed rules to curb improper market manipulation while tightening communication and reporting requirements for electric power and natural gas markets.

2006

Consumers burned more than four barrels of oil for every new barrel discovered. According to WREC by the year 2070, 70 percent of electricity worldwide would be produced by renewable resources.

July

A barrel of oil costs $78.40, 3 times more than in September 2003.

September

BMW unveils the first luxury car to be powered by liquid hydrogen.

Related Organizations

Alternative Fuels Data Center (AFDC)

1000 Independence Ave. SW

Washington, DC 20585

phone: (800) DIAL-DOE

fax: 202-586-4403

Web site: www.afdc.doe.gov

The Alternative Fuels Data Center has information on alternative-fuel vehicles and other advanced vehicles. It gives basic information on alternative fuels, with sections on hydrogen and alternative-fuel infrastructure, plus an interactive tool to compare the technical characteristics of fuels. AFDC publishes *Clean Cities Now* magazine and various documents and research papers concerning alternative fuels.

American Wind Energy Association (AWEA)

122 C St. NW, Suite 380

Washington, DC 20001

phone: (202) 383-2500

fax: (202) 383-2505

e-mail: windmail@awea.org

Web site: www.awea.org

The American Wind Energy Association is a national trade association that represents wind power plant developers, wind turbine manufacturers, utilities, and others involved in the wind energy industry. The AWEA promotes wind energy as a clean source of electricity for consumers around the world.

The Association for the Study of Peak Oil & Gas (ASPO)

Box 25182

SE-750 25 Uppsala

Sweden
phone: +46 70 4250604
e-mail: aleklett@tsl.uu.se
Web site: www.peakoil.net

The Association for the Study of Peak Oil & Gas is an international organization of scientists attempting to determine the date and impact of peak oil production and decline. The group publishes *ASPO,* a monthly newsletter.

Committee for a Constructive Tomorrow (CFACT)

PO Box 65722
Washington, DC 20035
phone: (202)429-2737
e-mail: info@cfact.org
Web site: www.cfact.org

The Committee for a Constructive Tomorrow supports continued development of atomic power and works to promote free-market and technological solutions to such growing concerns as energy production, food production and processing, air and water quality, wildlife protection, and much more. CFACT produces a national radio commentary called *Just the Facts* that is heard daily on some 300 stations across the United States.

Hubbert Peak of Oil Production

PO Box 7080
Santa Cruz, CA 95061
phone: (831) 425-8523
e-mail: webmaster@hubbertpeak.com
Web site: www.hubbertpeak.com

Named after the late geophysicist Marion King Hubbert, the organization provides data, analysis, and recommendations regarding the upcoming peak in the rate of global oil extraction.

National Biodiesel Board

3337a Emerald Lane
PO Box 104898
Jefferson City, MO 65110-4898
phone: (800) 841-5849
fax: (573) 635-7913
e-mail: info@biodiesel.org
Web site: www.biodiesel.org

The mission of the National Biodiesel Board is to create a sustainable biodiesel industry through public affairs, communications, technical, and quality assurance programs. The board believes that by 2015 biodiesel will be viewed as an integral component of a national energy policy that increasingly relies on clean, domestic, renewable fuels and that sales of biodiesel blends will exceed 1 billion gallons per year. The board publishes *Biodiesel* magazine, a newsletter, and various brochures available online.

The National Renewable Energy Laboratory (NREL)

1617 Cole Blvd.
Golden, CO 80401-3393
phone: (303) 275-3000
Web site: www.nrel.gov

The National Renewable Energy Laboratory is the U.S. Department of Energy's laboratory for renewable energy research, development, and deployment and a leading laboratory for energy efficiency. Some of the areas of scientific investigation at NREL include wind energy, biomass-derived fuels, advanced vehicles, solar manufacturing, hydrogen fuel cells, and waste-to-energy technologies. The organization publishes dozens of comprehensive research papers concerning these technologies, many of them available online for free.

Natural Resources Defense Council (NRDC)

40 W. 20th St.

New York, NY 10011
phone: (212) 727-2700
fax: (212) 727-1773
e-mail: nrcdinfo@nrdc.org
Web site: www.nrdc.org

The Natural Resources Defense Council uses law, science, and the support of 1.2 million members and online activists to protect the environment. The NRDC publishes *On Earth,* a quarterly magazine; *Nature's Voice,* a bimonthly online magazine; and various reports and e-mail bulletins.

Nuclear Energy Institute (NEI)

1776 I St. NW, Suite 400
Washington, DC 20006-3708
phone: (202) 739-8000
fax: (202) 785-4019
e-mail: webmasterp@nei.org
Web site: www.nei.org

The Nuclear Energy Institute is the policy organization of the nuclear energy industry, whose objective is to promote policies that benefit the nuclear energy business and to develop policy on key legislative and regulatory issues affecting the nuclear industry. The organization has over 260 corporate members in 15 countries, including companies that operate nuclear power plants, design and engineering firms, fuel suppliers and service companies, and companies involved in nuclear medicine and nuclear industrial applications. NEI publishes numerous books and brochures that promote nuclear energy and safety.

Renewable Energy Policy Project (REPP)

1612 K St. NW, Suite 202
Washington, DC 20006
phone: (202) 293-2898
fax: (202) 298-5857

e-mail: info2@repp.org

Web site: www.repp.org

The Renewable Energy Policy Project provides information about solar, hydrogen, biomass, wind, hydro, and other forms of "green" energy. The goal of the group is to accelerate the use of renewable energy by providing credible facts, policy analysis, and innovative strategies concerning renewables. REPP seeks to define growth strategies for renewables that respond to competitive energy markets and environmental needs. The project has a comprehensive online library of publications dedicated to these issues.

Sierra Club

85 Second St., 2nd Floor

San Francisco, CA 94105-3441

phone: (415) 977-5500

fax: (415) 977-5799

e-mail: information@sierraclub.org

Web site: www.sierraclub.org

The Sierra Club is a nonprofit public interest organization that promotes conservation of the natural environment by influencing public policy decisions—legislative, administrative, legal, and electoral. It publishes *Sierra* magazine as well as books on the environment.

U.S. Department of Energy

1000 Independence Ave. SW

Washington, DC 20585

phone: (800) DIAL-DOE

fax: (202) 586-4403

Web site: www.doe.gov

The Department of Energy's overarching mission is to advance the national, economic, and energy security of the United States; to promote scientific and technological innovation in support of that mission; and to ensure the environmental cleanup of the national nuclear weapons complex.

World Renewable Energy Congress (WREC)

PO Box 362

Brighton BN2 1YH

United Kingdom

phone: +44 1273 625643

e-mail: asayigh@netcomuk.co.uk

Web site: www.wrenuk.co.uk

World Renewable Energy Congress (WREC) is a major nonprofit organization registered in the United Kingdom and affiliated with UNESCO (the United Nations Educational, Scientific, and Cultural Organization). Established in 1992, WREC is one of the most effective organizations in supporting and enhancing the utilization and implementation of renewable energy sources that are both environmentally safe and economically sustainable. WREC publishes the journal *Renewable Energy* with articles on photovoltaic technology conversion, solar thermal applications, biomass conversion, wind energy technology, energy conservation in buildings, and other topics.

For Further Research

Books

Paula Berinstein, *Alternative Energy: Facts, Statistics, and Issues*. Westport, CT: Oryx, 2001.

Pierre Chomat, *Oil Addiction: The World in Peril*. Boca Raton FL: Universal, 2004.

Julian Darley, *High Noon for Natural Gas: The New Energy Crisis*. White River Junction, VT: Chelsea Green, 2004.

Seth Dunn, *Hydrogen Futures: Toward a Sustainable Energy System*. Washington, DC: Worldwatch Institute, 2001.

Howard S. Geller, *Energy Revolution: Policies for a Sustainable Future*. Washington, DC: Island, 2003.

Jeff Goodell, *Big Coal: The Dirty Secret Behind America's Energy Future*. Boston: Houghton Mifflin, 2006.

David Goodstein, *Out of Gas*. New York: Norton, 2004.

Scott W. Heaberlin, *A Case for Nuclear-Generated Electricity: (Or, Why I Think Nuclear Power Is Cool and Why It Is Important That You Think So Too)*. Columbus, OH: Battelle, 2004.

Richard Heinberg, *The Party's Over: Oil, War and the Fate of Industrial Societies*. Gabriola, BC: New Society, 2003.

Paul Kruger, *Alternative Energy Resources: The Quest for Sustainable Energy*. Hoboken, NJ: John Wiley & Sons, 2006.

Robert C. Morris, *The Environmental Case for Nuclear Power*. St. Paul, MN: Paragon, 2000.

Paul Roberts, *The End of Oil: On the Edge of a Perilous New World*. Boston: Houghton Mifflin, 2005.

Ian Rutledge, *Addicted to Oil: America's Relentless Drive for Energy Security*. New York: I.B. Tauris, 2005.

Vaclav Smil, *Energy at the Crossroads: Global Perspectives and Uncertainties.* Cambridge, MA: MIT Press, 2003.

Peter Tertzakian, *A Thousand Barrels a Second.* New York: McGraw-Hill, 2006.

Periodicals

Matthew H. Brown, "Energy Crisis Deja Vu: The U.S. Energy Needs—and the World's—Are Changing Fast. Are We Prepared?" *State Legislatures*, February 2005.

Ty Cashman, "The Hydrogen Economy," *Earth Island Journal*, Summer 2001.

Will Dana, "Al Gore 3.0," *Rolling Stone*, July 13–27, 2006.

Marla Dickerson, "Will Mexico Soon Be Tapped Out?" *Los Angeles Times*, July 24, 2006.

Manimoli Dinesh, "Hydrogen Fuel No Near-Term Panacea for Oil Dependence," *Oil Daily*, Feb 9, 2004.

Engineer, "Solar Power from Space: Sun Seekers," March 11, 2005.

Christopher Flavin, "Over the Peak," *World Watch*, January/February 2006.

Glenn Hamer, "Solar Power 2002," *World & I*, June 2002.

———, "Tipping Point: View on Renewables," *Power Engineering*, May 2003.

Valli Herman, "Vegetable Juice," *Los Angeles Times*, August 5, 2006.

Peter Maass, "The Breaking Point," *New York Times*, August 21, 2005.

Charles J. Murray, "Fuel Cell R&D Is Far from Easy Street," *Electronic Engineering Times*, May 26, 2003.

Barack Obama, "Fueling the Future," *American Prospect*, April 2006.

Mike Oliver and John Hospers, "Alternative Fuels?" *American Enterprise*, September 2001.

Greg Pahl, "Heat Your Home with Biodiesel," *Mother Earth News*, December 2003–January 2004.

Joe Provey, "The Sun Also Rises—Again: After Decades of Setbacks, Solar Energy Returns and Many Believe That It's Here to Stay," *Popular Mechanics,* September 2002.

Jeremy Rifkin, "Hydrogen: Empowering the People; A New Source of Energy, If Its Development Follows the Model of the World Wide Web, Offers a Way to Wrench Power from Ever Fewer Institutional Hands," *Nation*, December 23, 2002.

Paul Roberts, "Running Out of Oil—and Time," *Los Angeles Times,* March 7, 2004.

Union of Concerned Scientists, "Energy Security: Solutions to Protect America's Power Supply and Reduce Oil Dependence," Cambridge, MA: UCS Publications, 2002.

Bryant Urstadt, "Imagine There's No Oil," *Harpers,* August 2006.

Internet Sources

Ronald Bailey, "Moonshine Mirage," *Reason Online*, May 12, 2006. www.reason.com/rb/rb051206.shtml.

Paul K. Driessen, "The False Promise of Renewable Energy," The Heartland Institute, May 2001. www.heartland.org/Article.cfm?artId=1122.

National Biodiesel Board, "Biodiesel," 2006. www.biodiesel.org/resources/faqs/default.shtm.

Nuclear Management Company, "Nuclear Energy and the Environment," July 2000. www.nmcco.com/education/facts/environment/energy.htm.

David J. O'Reilly, "Will You Join Us," willyoujoinus.com, 2006. http://www.willyoujoinus.com.

Robert D. Perlack, Lynn L. Wright, Anthony F. Turhollow et al. "Biomass as Feedstock for Bioenergy and Bioproducts Industry: The Technical Feasibility of a Billion-Ton Supply," U.S. Department of Energy, April 2005. http://feedstockreview.ornl.gov/pdf/billion_ton_vision.pdf.

Project on Government Oversight, "Nuclear Power Plant Security: Voices from Inside the Fences," September 12, 2002. www.pogo.org/p/environment/eo-020901-nukepower.html#ExecSum.

Stuart H. Rodman, "When Drill Holes Become Rat Holes," Ecological Life Systems Institute, 2000. www.elsi.org/endofoil.htm.

Walter Youngquist, "Alternative Energy Sources," The Coming Global Oil Crisis, October 2000. www.oilcrisis.com/youngquist/altenergy.htm.

Source Notes

Overview

1. Peter Tertzakian, *A Thousand Barrels a Second*. New York: McGraw-Hill, 2006, p. 1.
2. Walter Youngquist, "Alternative Energy Sources," *The Coming Global Oil Crisis*, October 2000, www.oilcrisis.com.
3. Quoted in Peter Maass, "The Breaking Point," *New York Times*, August 21, 2005, p. 3.
4. Julian Darley, *High Noon for Natural Gas*. White River Junction, VT: Chelsea Green, 2004, pp. 8–9.
5. Jeff Goodell, *Big Coal: The Dirty Secret Behind America's Energy Future*. Boston: Houghton Mifflin, 2006, p. xv.
6. Goodell, *Big Coal*, p. xv.
7. Youngquist, "Alternative Energy Sources."

Is the World Running Out of Oil?

8. Tertzakian, *A Thousand Barrels a Second*, p 3.
9. Maass, "The Breaking Point," p 3.
10. David Deming, "Abundant Reserves Show Petroleum Age Is Just Beginning," Heartland Institute, October 2003. www.heartland.org.
11. Quoted in Will Dana, "Al Gore 3.0," *Rolling Stone*, July 13–27, 2006, p. 46.

Will Alternative-Fuel Vehicles Solve the World Energy Crisis?

12. Ronald Bailey, "Moonshine Mirage," *Reason Online*, May 12, 2006. www.reason.com.
13. Quoted in Robert Bryce, "Corn Dog: The Ethanol Subsidy Is Worse than You Can Imagine," *Slate.com*, July 19, 2005. www.slate.com.

14. Jules Dervaes, "Projects: Backyard Biodiesel," Path to Freedom, March 9, 2005. www.pathtofreedom.com.
15. Jeremy Rifkin, "Hydrogen: Empowering the People," *Nation*, December 23, 2002, p. 20.

Can Nuclear Power Supply the World's Energy Needs?

16. Quoted in Reuters, "After Gas Crisis, Germany Takes Another Look at Nuclear Power," January 4, 2006. www.iht.com.
17. Nuclear Energy Institute, "Environmental Preservation," 2004. www.nei.org.
18. Richard Black, "Finland Buries Its Nuclear Past," BBC News, April 27, 2006. http://news.bbc.co.uk.
19. Quoted in Black, "Finland Buries Its Nuclear Past."
20. Greenpeace, "End the Nuclear Age," May 30, 2006. www.greenpeace.org.

Can Renewable Energy Meet Future Demands for Electricity?

21. Quoted in Godfrey Boyle, ed. *Renewable Energy: Power for a Sustainable Future*. New York: Oxford University Press, 1996, p. 27.
22. Quoted in Geir Moulson, "Nations Vow to Promote Renewable Energy," Health and Energy, June 4, 2004. http://healthandenergy.com.
23. Quoted in Stuart H. Rodman, "When Drill Holes Become Rat Holes," Ecological Life Systems Institute, 2000. www.elsi.org.
24. Quoted in "DuPont Executive Outlines Company's Plan for Growth in Alternative Energy Technologies," *PR News Today*, May 16, 2006. www.

prnewstoday.com.

25. Eric Rosenbloom, "A Problem with Wind Power," January 2004. www.kirbymountain.com.

26. Rosenbloom, "A Problem with Wind Power."

27. Mike Oliver and John Hospers, "Alternative Fuels?" *American Enterprise*, September 2001, p. 20.

28. Paul Carlstrom, "As Solar Gets Smaller, Its Future Gets Brighter," *San Francisco Chronicle*, July 11, 2005. www.sfgate.com.

List of Illustrations

Index

About the Author

Stuart A. Kallen is a prolific author who has written more than 200 non-fiction books for children and young adults over the past 20 years. His books have covered countless aspects of human history, culture, and science from the building of the pyramids to the music of the 21st century. Some of his recent titles include *History of World Music, Romantic Art,* and *Women of the Civil Rights Movement.* Kallen is also an accomplished singer-songwriter and guitarist in San Diego, California.